## Author

Debbie Snyder has been an educator and former homeschool mom of seven children for over forty years. Her teaching includes Bible, science, anatomy, health, speech, and debate in public schools, Christian schools, and homeschool co-ops. Her goal is to pass the torch to the next generation.

## Illustrator and Advisor

Jeanette Summerville is a registered nurse, a former homeschool mom of eight children, an artist, and a dedicated missionary to the Deaf of Guatemala since 2010. With over 35 years of experience teaching and creating curriculum for both Sunday school and for the Deaf, she is passionate about making learning meaningful and accessible for all ages.

## In Loving Memory
**of**
## William Baker Snyder, Jr.
**(1950-2021)**

## Dedicated
**to**
## Snyder and Summerville Children/Cousins
David, John, Tom, Tim, Anna, Sarah, and Becca
Esther, Liz, Hannah, Nathan, Lydia, Josh, Luke, and Julia
And all of our grandchildren.

## Target Audience Tests
(Tested in-person and online)
Berryville Homeschool Academy: Berryville, VA (ages 7-9)
Classical Cottage School: Winchester, VA (ages 5-9)
Grandchildren: In-person and online (ages 3-9)
Guatemala Deaf Ministries: El Progreso, Guatemala (ages 5-14)

## How to Use
Each book represents a day of creation found in Genesis 1 and 2.
Pick and choose activities according to abilities and interests.
As a unit study, this material is great for teaching
different ages at the same time.

Exploring With Grandma Publishing, Virginia
exploringwithgrandma@gmail.com
exploringwithgrandma.com

Complete list of Images and Sources available in *Exploring God's Creation Teacher's Helps* (companion book).
Edited by Jeanette Summerville, Carrie Tucker, and Irina Snyder

# Table of Contents

# Introduction
## Creation

*In the beginning, God created the heavens and the earth.*
*(Genesis 1:1)*

**Day 1**
**Day and Night**

**Day 2**
**Sky and Water**

**Day 3**
**Dry Land and Plants**

**Day 4**
**Sun, Moon, and Stars**

**Day 5**
**Fish and Birds**

**Day 6**
**Animals and People**

**Day 7**
**God Rested**

## Creation Song

Tune of "Brother John"

God created, God created,
Day and night, day and night.
That was on the first day.
That was on the first day.
It was good. It was good.

God created, God created,
Sky and water, sky and water.
That was on the second day.
That was on the second day.
It was good. It was good.

God created, God created,
Land and plants, land and plants.
That was on the third day.
That was on the third day.
It was good. It was good.

God created, God created,
Sun, moon, and stars,
Sun, moon, and stars.
That was on the fourth day.
That was on the fourth day.
It was good. It was good.

God created, God created,
Fish and birds,
Fish and birds.
That was on the fifth day.
That was on the fifth day.
It was good. It was good.

God created, God created,
Animals and people,
Animals and people.
That was on the sixth day.
That was on the sixth day.
It was good. It was good.

Then God rested, then God rested,
From His work, from His work.
That was on the seventh day.
That was on the seventh day.
It was good. It was good.

Let's sing praises. Let's sing praises,
To our God, to our God.
He is everywhere,
Giving love and care.
God is good. God is good.

# Creation
## Adam and Eve

God created a beautiful garden for Adam and Eve. Trees, vines, and bushes filled the earth with tasty fruits and vegetables for them to eat. *The Lord God commanded the man, saying, "From any tree of the garden you may freely eat; but from the tree of the knowledge of good and evil you shall not eat, for on the day that you eat from it you will certainly die." (Genesis 2:16-17)*

For a while, Adam and Eve obeyed God. Their lives were peaceful and happy. Then one fateful day, Satan slithered into the garden in the form of a beautiful serpent. The serpent lied and told Adam and Eve they would not die if they ate the forbidden fruit. Instead, they would become like God, knowing good and evil. They ate the fruit and sinned.

As a consequence, Adam and Eve had to leave the beautiful garden. Sin and suffering entered the world. Life became more difficult. Eventually, Adam and Eve died.

God was still in control and had a plan. He promised there would be a day when He would send His son, Jesus, to crush Satan's power forever. The world will be restored to its original beauty once again.

### Chain Reaction

1. What was the cause?
2. What was the effect?

# Sin
## Cain and Abel

Cain and Abel were sons of Adam and Eve. Cain was a farmer and Abel was a shepherd. In the course of time, the two of them brought sacrifices to God. Cain offered fruit from the ground; Abel offered a lamb. When God was pleased with Abel's offering and displeased with Cain's offering, Cain became furious.

Then the Lord said to Cain, *"Why are you angry? And why is your face gloomy? If you do well, will your face not be cheerful? And if you do not do well, sin is lurking at the door; and its desire is for you, but you must master it." (Genesis 4:6-7)*

God gave an opportunity for Cain to do the right thing, but Cain still chose to do evil. Filled with anger, he cruelly killed his brother in a field.

God was merciful and put a mark on Cain to warn people not to kill him. He became a wanderer on the earth. He could no longer grow crops. (Genesis 4)

### Chain Reaction

1. What was the cause?
2. What was the effect?

## Rescue
### Jesus

*For the Lord Himself will descend from heaven with a shout,*
*with the voice of the archangel and with the trumpet of God,*
*and the dead in Christ will rise first.*
*Then we who are alive, who remain,*
*will be caught up together with them in the clouds to meet the Lord in the air,*
*and so we will always be with the Lord.*
*Therefore, comfort one another with these words.*
*(I Thessalonians 4:16-18)*

Do you remember God's plan to crush Satan's power forever? It took many years, but it finally happened! Jesus, God's only Son, came to the earth as a baby. He died on a cross as a perfect and innocent sacrifice, so our sins could be forgiven.

All people die, but Jesus' death was different. After being dead for three days, He came alive again! For the next forty days, He taught about God's kingdom to hundreds of people.

As people watched Jesus rise into the sky towards heaven, Jesus promised He would come back again. His followers are still faithfully watching and waiting for this to happen.

When we trust in Jesus' sacrifice on the cross, admitting all the wrong things we have done, Jesus will forgive our sins and declare us innocent and clean before God. Someday Jesus will return. You can count on it!

### Chain Reaction

1. What was the cause?
2. What was the effect?

# CREATION
# WORD SEARCH

| | | | | | | | | | | | | | |
|---|---|---|---|---|---|---|---|---|---|---|---|---|---|
| N | A | J | W | A | T | S | H | L | A | G | R | O | T |
| O | P | B | U | K | Y | L | A | N | D | B | N | F | C |
| I | Z | L | V | U | O | T | S | L | M | S | L | A | F |
| T | I | C | A | R | M | O | O | N | K | K | E | D | P |
| A | F | D | E | N | D | E | C | H | S | Y | T | C | R |
| E | H | N | E | P | T | W | E | I | B | A | D | E | H |
| R | G | U | A | F | G | C | N | I | J | U | T | L | A |
| C | F | O | V | M | D | O | U | I | B | A | G | N | T |
| V | U | E | W | K | U | E | S | J | W | O | I | X | I |
| A | D | B | S | C | V | N | I | M | D | G | G | E | H |
| U | O | A | O | T | L | U | O | A | H | W | A | O | I |
| K | O | X | Y | N | A | M | P | T | S | I | M | E | O |
| Y | P | U | I | Y | I | R | A | T | X | A | O | U | E |
| A | N | I | M | A | L | A | Z | I | R | E | Q | H | I |

| | | |
|---|---|---|
| LAND | MOON | WATER |
| DAY | ANIMAL | CREATION |
| NIGHT | SKY | PLANT |
| SUN | STAR | MAN |

# WORD SEARCH

Make Your Own

## Chain Reactions

*By the word of the Lord the heavens were made,*
*And by the breath of His mouth all their lights.*
*For He spoke, and it was done;*
*He commanded, and it stood firm.*
*(Psalm 33:6, 9)*

### Vocabulary

**Cause:** when an object makes something happen

**Chain reaction:** when one action causes another action

**Effect:** a consequence

### Cause and Effect

# Domino Chain Reaction
## Activity

**Purpose**
To demonstrate a chain reaction (cause and effect)

**Materials**
Dominoes

**Procedure**
Set dominoes vertically upright in a row (not horizontal).
Slightly knock over the first domino.

**Result**
A chain reaction will occur.

**Why?**
A chain reaction happens when one action causes another action.

## A Firm Foundation

*He is like a man building a house, who dug deep and laid a foundation on the rock;*
*and when there was a flood, the river burst against that house and*
*yet it could not shake it, because it had been well built.*
*(Luke 6:48)*

### Vocabulary

**Cornerstone:** a large stone used in ancient times during the construction of a building

**Foundation:** the supporting materials underneath a building (stone or sand)

**Horizontal:** east/west; left/right; where the earth appears to meet the sky (horizon)

**Plumb:** vertical and straight

**Plumb line**: a line with a weight on one end used to determine if something is vertically straight

**Vertical:** up/down; north/south

## A Firm Foundation

*Therefore, everyone who hears these words of Mine, and acts on them, will be like a wise man who built his house on the rock. And the rain fell and the floods came, and the winds blew and slammed against that house; and yet it did not fall, for it had been founded on the rock.*

*And everyone who hears these words of Mine, and does not act on them, will be like a foolish man who built his house on the sand. And the rain fell and the floods came, and the winds blew and slammed against that house; and it fell—and its collapse was great.*
*(Matthew 7:24-27)*

## The Wise Man Built His House Upon a Rock

The wise man built his house upon the rock.
The wise man built his house upon the rock.
The wise man built his house upon the rock.
And the rain came tumbling down.

The rain came down, and the floods came up.
The rain came down, and the floods came up.
The rain came down, and the floods came up.
And the house on the rock stood firm!

The foolish man built his house upon the sand.
The foolish man built his house upon the sand.
The foolish man built his house upon the sand.
And the rain came tumbling down.

The rain came down, and the floods came up.
The rain came down, and the floods came up.
The rain came down, and the floods came up.
And the house on the sand went SPLAT!

# Wooden Blocks
## Activity

**Purpose**
To illustrate the importance of a good foundation

**Materials**
Wooden blocks

**Procedure**
Stack blocks in a crisscross fashion.
Take turns and carefully pull one block at a time from the structure.
The person who causes the structure to fall by removing the wrong block loses.

**Result**
The structure falls when the wrong piece is removed.

**Why?**
The foundation gets weaker and weaker as more and more pieces are removed.

# Plumb line
## Activity

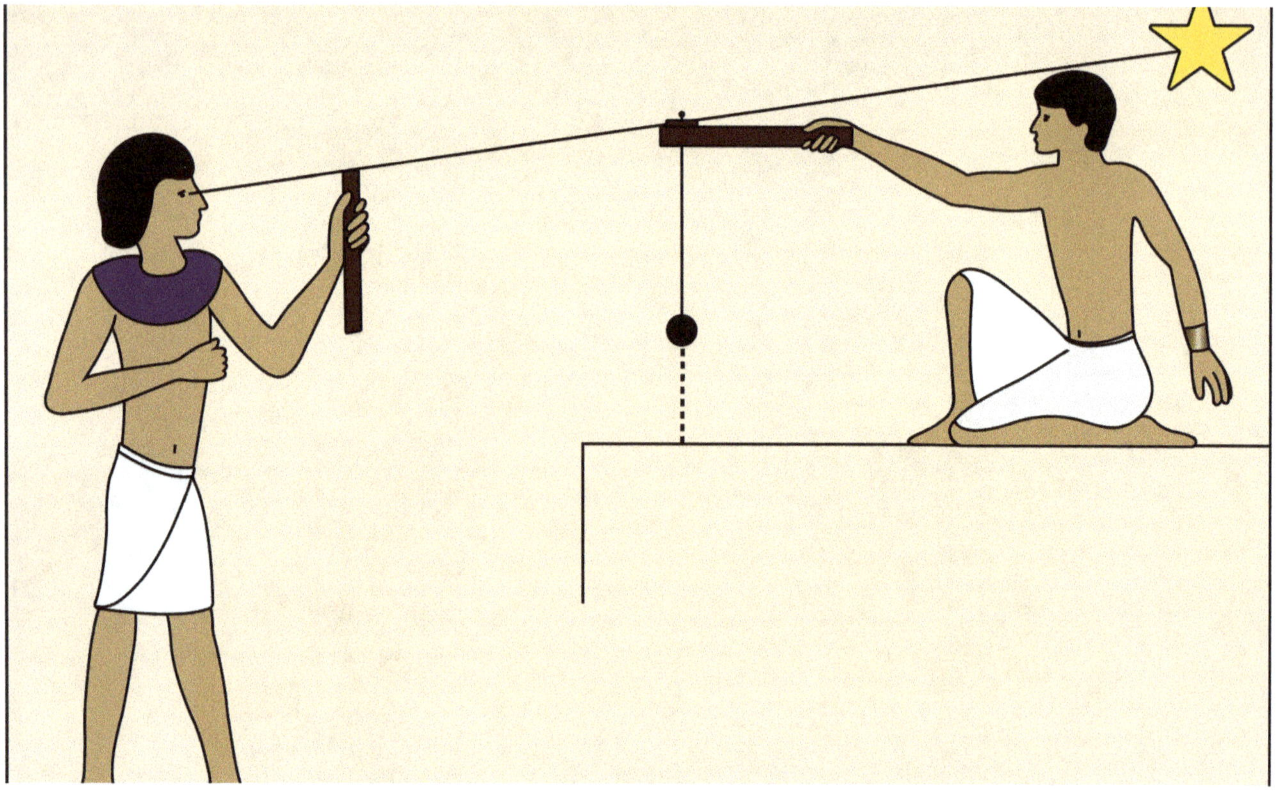

*Plumb lines were used to build Egyptian pyramids.*

**Purpose**
To create a plumb line

**Materials**
10-inch string
½-inch metal washer
Tack with a knob
Picture

**Procedure**
Tie a tack to one end of the string.
Tie a metal washer to the other end.
Stick the tack into the wall next to a picture.
Let the plumb line hang freely, approximately one inch from the picture.

**Result**
The plumb line will help determine whether or not the picture is straight.

**Why?**
A plumb line provides a standard. All work must line with the plumb line or risk being crooked. Plumb lines can be used to build bridges, install windowpanes, and hang pictures.

## Plumb line
### Spiritual Activity

*I will make justice the measuring line and righteousness the plumb line…*
*(Isaiah 28:17)*

**Purpose**
To acquire life application with a plumb line

**Materials**
Two Paths: Coloring sheet (page 19)
Colored pencils, markers, or crayons

**Procedure**
Color the narrow path leading to God (plumb).
Leave the wicked side black and white.

**Result**
The narrow path leads to life. The wide path leads to destruction.

**Why?**
The words of the Bible are a type of plumb line. The Bible does not change. Instead, it is a standard, helping Christians to live righteous lives in this ever-changing, evil world.

# Two Paths

*Enter through the narrow gate; for the gate is wide and the way is broad*
*that leads to destruction, and there are many who enter through it.*
*For the gate is narrow and the way is constricted that leads to life,*
*and there are few who find it.*
*(Matthew 7:13-14)*

## Cornerstone
### Activity

*[Jesus] is the stone which was rejected by you, the builders, but which became the chief cornerstone. And there is salvation in no one else; for there is no other name under heaven that has been given among mankind by which we must be saved.*
*(Acts 4:11-12)*

Copy, trace, or color the cornerstone.

In ancient times, builders used cornerstones for construction projects. Once the cornerstone was set, it became the basis for determining every measurement in the remaining construction; everything was aligned to the cornerstone.

In the same way, the Bible describes Jesus as the cornerstone of the church. As the cornerstone, Jesus becomes our standard of measurement and alignment. We look to Jesus' life and teachings to determine how to live.

# A Firm Foundation
## The Three Little Pigs

The first pig built his house out of straw.
The second pig built his house out of sticks.
The third pig built his house out of bricks.
Which pig was safe from the wolf?

### Questions: A Firm Foundation

1. If a line goes north to south, it is ______________________________.

2. If a line goes east to west, it is ______________________________.

3. An instrument used to determine if something is vertically straight is called a

______________________ ______________________.

4. If a picture is vertically straight, it is considered ______________________.

5. In ancient times, the stone placed at the corner of a building under construction was called a

______________________.

6. The supporting material underneath a building is called a ______________________.

# Unit 1
## Day and Night

*I once was blind but now I see…*
John Newton (former slave trader)

## Day 1
## Day and Night

*And God said, "Let there be light," and there was light. And God saw that the light was good. And God separated the light from the darkness. God called the light Day, and the darkness He called Night. And there was evening and there was morning, the first day.*
*(Genesis 1:3-5)*

# Chapter 1
## Reflective and Refractive Light

*But if we walk in the light, as He is in the light, we have fellowship with one another, and the blood of Jesus, His Son, purifies us from all sin. (I John 1:7)*

### Vocabulary

**Reflective light:** light bouncing off an object (glass, water, or polished metal)

**Refractive light:** light bending (magnifying glass, prism, or rainbow)

### Reflective Light

### Refractive Light

# Reflective Light
## Activity

**Purpose**
To learn about reflective light

**Materials**
3x5 index cards
Mirror
Pencil

**Procedure**
Write your name on a card.
Place the card in front of a mirror.

**Result**
Your name will appear backwards.

**Why?**
The image of the letters reflects backwards retracing the path it traveled, reversing the letters.

# Refractive Light
## Activity

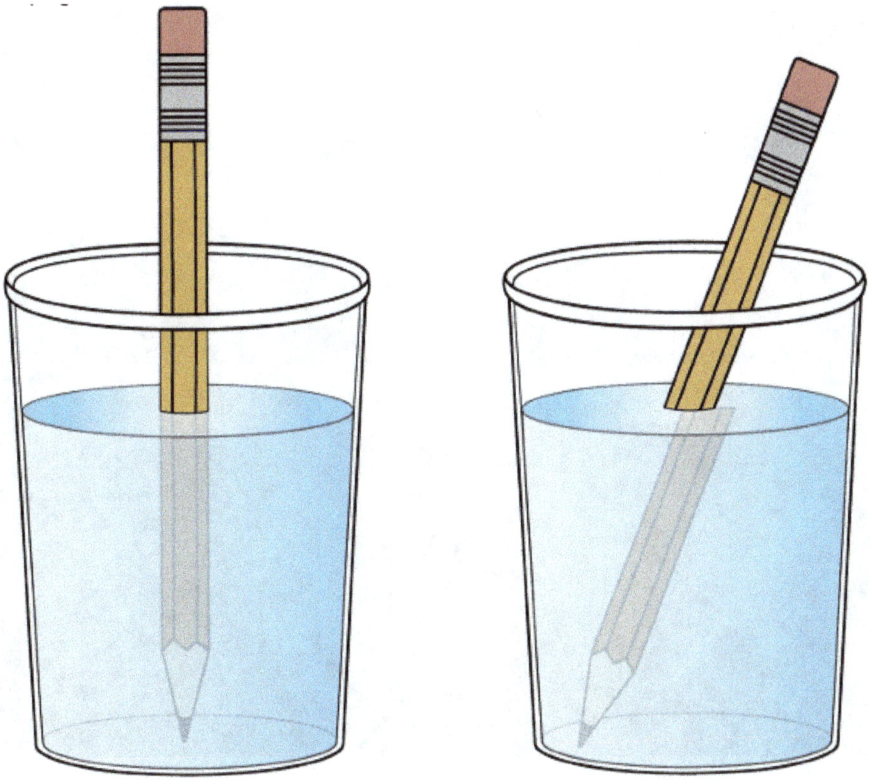

**Purpose**
To observe light as it bends when it passes from one material to another

**Materials**
Glass jar
Water
Pencil

**Procedure**
Fill the jar with water.
Place a pencil sideways in the jar.

**Result**
The pencil will appear to break into two pieces.

**Why?**
When light bends, our eyes see optical illusions.

**Circle the correct answer.**

1. Light bends. (reflection or refraction)
2. Light bounces off an object. (reflection or refraction)

# Reflective and Refractive Light
## Activity

Discover reflective and refractive light around your home. Draw pictures or write words.

**Reflective light:** light bouncing off an object (glass, water, or polished metal)

1.

2.

3.

4.

5.

**Refractive light:** light bending (magnifying glass, prism, or rainbow)

1.

2.

3.

4.

5.

# Chapter 2
## Colors

### Flood

Noah was a righteous man who lived during a time when the earth was filled with wickedness. As punishment, God decided to destroy the entire earth with a flood. To save his family from drowning, Noah built an ark according to God's instructions.

When the ark was completed, Noah brought his family on board, along with a male and female of each kind of animal. It rained for forty days and forty nights. Water covered the whole earth. Everyone outside the ark drowned.

After many days, the water levels lowered, and land could be seen. God placed a rainbow in the sky as a sign that He would never again destroy the whole earth with a flood. (Genesis 6-9)

## Chain Reaction

1. What was the cause?
2. What was the effect?

## Rainbow

*God said, "...I have set My rainbow in the cloud, and it shall serve as a sign of a covenant between Me and the earth. It shall come about, when I make a cloud appear over the earth, that the rainbow will be seen in the cloud, and I will remember My covenant, which is between Me and you and every living creature of all flesh; and never again shall the water become a flood to destroy all flesh. When the rainbow is in the cloud, then I will look at it, to remember the everlasting covenant between God and every living creature of all flesh that is on the earth."*
*(Genesis 9:13-16)*

## Chain Reaction

1. What was the cause?
2. What was the effect?

# Rainbow
## Activity

**Purpose**
To recite the colors of the rainbow in order
To observe the circular nature of a rainbow
To connect rainbows with refractive light

**Materials**
Colored pencils, markers, or crayons
Rainbow coloring page (page 31)

**Procedure**
Color each band of the rainbow as listed.

**Result**
The rainbow is refractive and circular in nature. Its colors are always in the same order.

**Why?**
The combination of light waves results in predictable colors for every rainbow.

# Rainbow

Color each band of the rainbow.

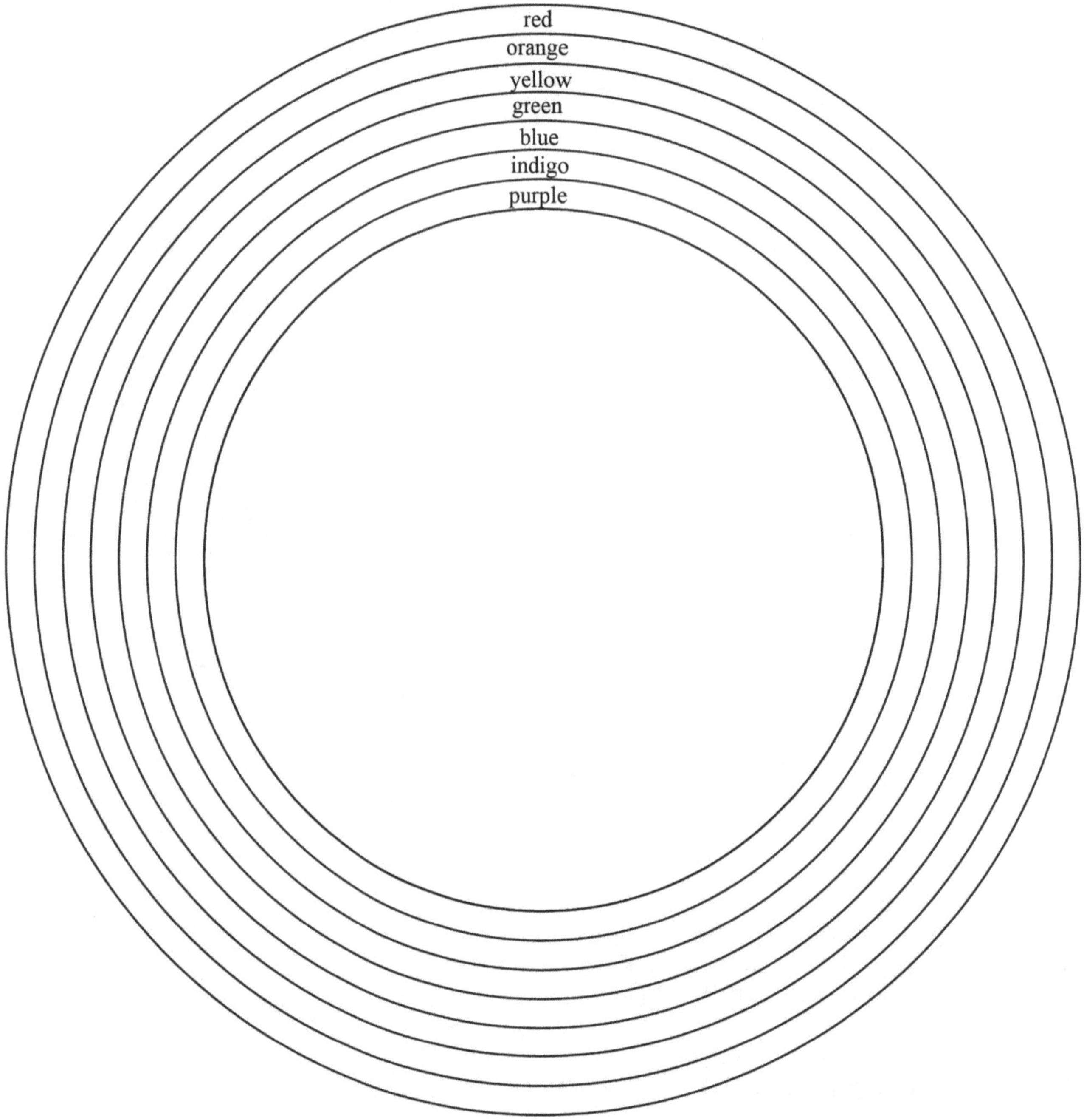


*No matter where you move,*
*a rainbow will always be the same distance away from you.*
*You can never reach the end of a rainbow.*

# COLORS IN THE BIBLE

Unscramble the words.

1. **ENGRE represents SESNFERSH and ERBRIHT.**

*But as for me, I am like a green olive tree in the house of God;*
*I trust in the faithfulness of God forever and ever.*
*(Psalm 52:8)*

*He lets me lie down in green pastures;*
*He leads me beside quiet waters.*
*(Psalm 23:2)*

2. **EWIHT signifies PYURIT, righteousness, and OHILESNS.**

*[Jesus'] appearance was like lightning, and His clothing as white as snow.*
*(Matthew 28:3)*

3. **DER represents DOLBO and NIS.**

*"Come now, and let us debate your case,"*
*Says the Lord,*
*"Though your sins are as scarlet,*
*They shall become as white as snow;*
*Though they are red like crimson,*
*They shall be like wool.*
*(Isaiah 1:18)*
*note: crimson and scarlet both mean a deep red color*

4. **ELPRPU represents CHRI and RYTOLAY. (King to Daniel)**

*But I personally have heard about [Daniel], that you are able to give interpretations and solve difficult problems. Now if you are able to read the inscription and make its interpretation known to me, you will be clothed with purple and wear a necklace of gold around your neck, and you will have authority as the third ruler in the kingdom."*
*(Daniel 5:16)*

5. **DGOL and RISLVE represents WHLAET.**

*And after they came into the house, they saw the Child with His mother Mary; and they fell down and worshiped Him. Then they opened their treasures and presented to Him gifts of gold, frankincense, and myrrh. (Matthew 2:11)*

*A word fitly spoken is like apples of gold in a setting of silver.*
*(Proverbs 25:11)*

1. ______________ ______________ ______________

2. ______________ ______________ ______________

3. ______________ ______________ ______________

4. ______________ ______________ ______________

5. ______________ ______________ ______________

# Colors
## Activity

### Vocabulary

**Primary colors:** blue, red, and yellow

**Secondary colors:** green, orange, and purple

**Purpose**
To observe what happens when colors are mixed

**Materials**
4 clear glasses
Food coloring (blue, red, and yellow)
Water

**Procedure**
Fill four glasses with water.

**First glass:** Add five drops each of red and yellow food coloring.

**Second glass:** Add five drops each of red and blue food coloring.

**Third glass:** Add five drops each of blue and yellow food coloring.

**Fourth glass:** Add five drops each of red, yellow, and blue food coloring.

**Result**
Red and yellow = orange
Red and blue = purple
Blue and yellow = green
Red, yellow, and blue = brown

**Why?**
The combination of different light wavelengths creates different colors.

## Color Wheel
### Activity

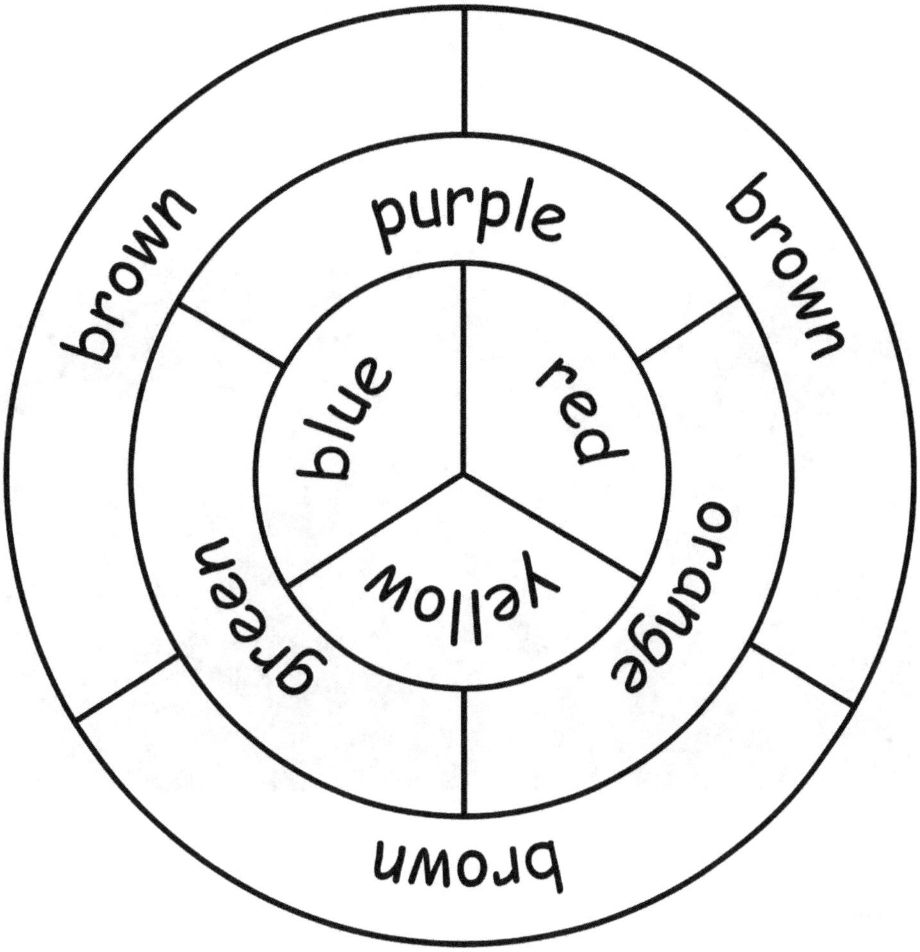


**Purpose**
To determine what happens when colors are mixed

**Materials**
Color wheel
Crayons

**Procedure**
Color the Color Wheel according to the words listed.

**Result**
Red and yellow = orange
Red and blue = purple
Blue and yellow = green
Orange and purple = a type of brown
Purple and green = a type of brown
Green and orange = a type of brown

**Why?**
When combining colors, the combination of light waves result in predictable colors.

# Colors

## Questions: Colors

1. What are the three primary colors?
2. What are the three secondary colors?
3. What are the seven colors of a rainbow?
4. Is a rainbow an arc or a circle?
5. What did God promise with the first rainbow?
6. Can you reach the end of a rainbow?
7. Is a rainbow an example of reflective or refractive light?

# Chapter 3
## Light and Darkness

*...nor do people light a lamp and put it under a basket, but on the lampstand, and it gives light to all who are in the house. Your light must shine before people in such a way that they may see your good works, and glorify your Father who is in heaven.*
*Matthew 5:5-6)*

### Vocabulary

**Bioluminescence:** the light coming from the bodies of some plants and animals

**Carnivore:** a meat-eating animal

**Darkness:** the lack of light

**Herbivore:** a plant-eating animal

**Insect:** an animal with six legs, exoskeleton, three body parts, two antennae, and wings

**Invertebrate:** an animal without a backbone

**Light:** a type of energy allowing sight

**Omnivore:** a plant and meat-eating animal

**Predator:** an animal hunting another animal

**Prey:** an animal hunted by another animal

**Vertebrate:** an animal with a backbone

## Light and Darkness
Spiritual Activity

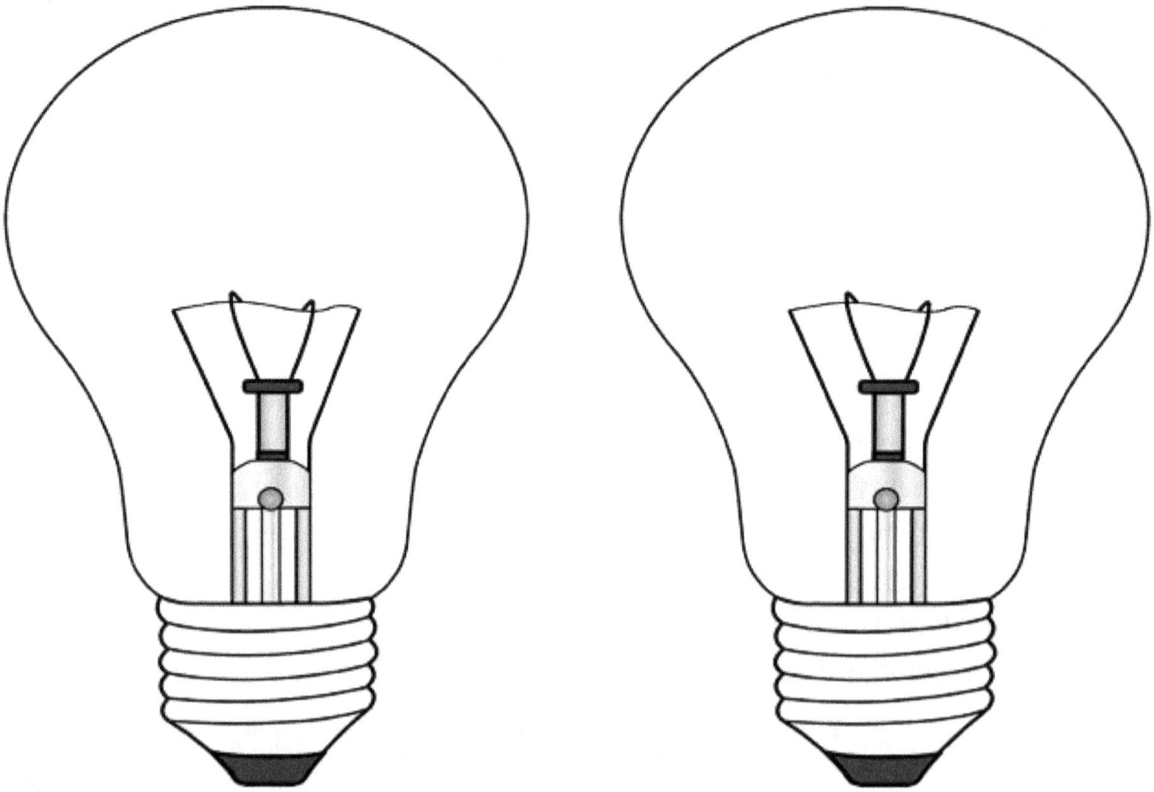

**Purpose**
To understand sin as a light-blocker
To appreciate Jesus as the light of the world

**Materials**
Print this page.
Glue
Black sand
Yellow colored pencils, markers, or crayons

**Procedure**
Color both light bulbs yellow.
Cover the first light bulb with glue.
Then sprinkle black sand on the glue.
(The black sand represents sin.)
Leave the second light bulb yellow.

**Result**
Sin blocks the light of Jesus.

**Why?**
*If we confess our sins, He is faithful and just to forgive our sins and cleanse us from all unrighteousness.*
*(I John 1:9)*

## Jesus Is the Light of the World.

*Then Jesus again spoke to them, saying, "I am the Light of the world; the one who followsMe will not walk in the darkness, but will have the Light of life."*
*(John 8:12)*

The light of Jesus shows us our sin. When we confess our wrong actions and thoughts, Jesus takes our sins upon Himself making it possible for us to stand clean and blameless before a holy God. We are forgiven.

## Questions

1. What word means no light at all?
2. What makes things visible?
3. Who is the light of the world?

# Fireflies

Male fireflies, also called lightning bugs, light up the night sky by bioluminescence, a natural light coming from their bodies. Also using bioluminescence, female fireflies shine light on the ground trying to attract male fireflies. The lights of fireflies can be green, pale red, or yellow.

Fireflies live in grassy areas, marshes, places with water, and woodlands. They eat smaller insects. God gave them a bad-tasting substance in their blood to discourage predators from tasting them. However, birds, other fireflies, and reptiles still eat them.

**Firefly**

Draw the other half of the firefly in the squares below.

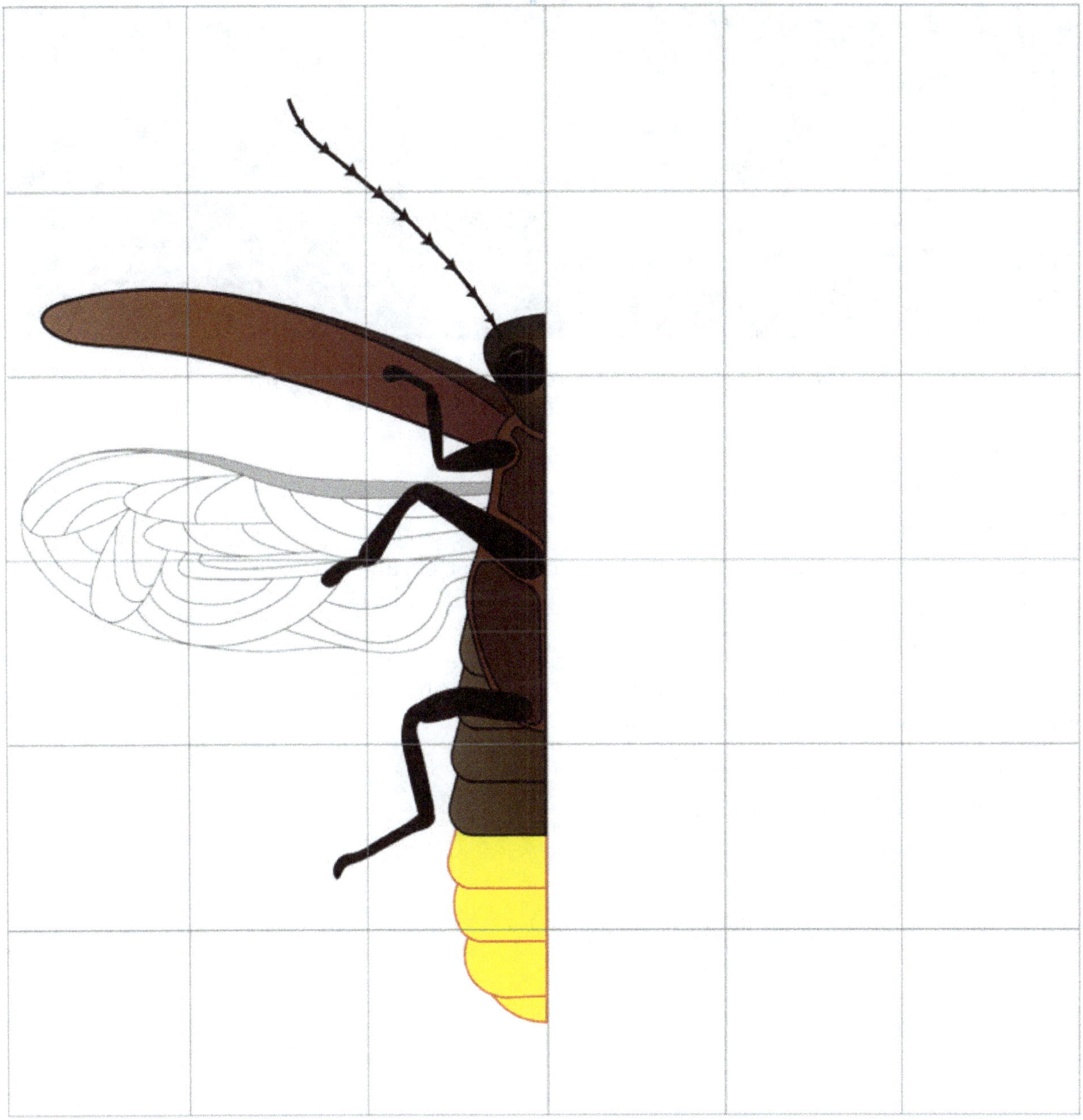

## Fireflies

*Have you ever tried to catch fireflies in jars at night?*
*Do you have a firefly story?*

### Questions: Fireflies

1. What is another name for fireflies?
2. Since a firefly is an insect, is it a vertebrate or an invertebrate?
3. Where do fireflies live?
4. What do fireflies eat?
5. If a firefly eats smaller insects, is it a carnivore, herbivore, or omnivore?
6. What are some of the predators of fireflies?
7. What protection did God give fireflies?
8. What colors do fireflies produce?
9. What is bioluminescence?
10. When is the best time to see a firefly?
11. Do male or female fireflies light up the night sky

# Jellyfish

Jellyfish are not fish, because they do not have vertebrates. They float around ocean currents eating small fish and zooplankton. Tentacles containing venom keep them safe from some predators. Larger jellyfish, sea turtles, sharks, penguins, and some fish still eat them. Ninety-five percent of a jellyfish's body is made of water.They live about one year.

Like the firefly, jellyfish produce a light called bioluminescence. The colors of a jellyfish include blue, lilac, orange, purple, red, and yellow. Some are even clear.

## Immortal Jellyfish

*If injured or threatened, an "immortal" jellyfish can go back to an earlier stage of development. These jellyfish can live a lot longer than one year.*

# Jellyfish

## Questions: Jellyfish

1. Are jellyfish fish?
2. Are jellyfish vertebrates or invertebrates?
3. Is a jellyfish a carnivore, herbivore, or omnivore?
4. What do jellyfish eat?
5. Where do jellyfish live?
6. What are the predators of jellyfish?
7. What protection did God give jellyfish?
8. What gives jellyfish light?
9. What colors are jellyfish?
10. How long do jellyfish live?
11. Jellyfish are mostly made up of what substance?

# Light and Darkness
## Activity

List three things in nature God designed to give light.

1.

2.

3.

List three things in the world humans design to give light.

1.

2.

3.

# Chapter 4
## Caves, Brown Bears, and Bats

*Then God said, "Let there be light"; and there was light. And God saw the light, that it was good; and God divided the light from the darkness. God called the light Day, and the darkness He called Night. So, the evening and the morning were the first day.*
*(Genesis 1:3-5)*

### Vocabulary

**Carnivore:** a meat-eating animal

**Cave:** a natural, underground hole that is completely dark

**Echolocation:** bouncing sounds off insects to locate them

**Herbivore:** a plant-eating animal

**Hibernate:** a deep sleep some animals take in order to get through the winter

**Mammal:** a warm-blooded vertebrate that has hair and nurses; most give birth to live young

**Omnivore:** a plant and meat-eating animal

**Stalactites:** dripping mineral water growing from cave ceilings into rock structures
*(Helpful hint: Stalactites stick tight to ceilings.)*

**Stalagmites:** dripping mineral water growing from cave floors into rock structures
*(Helpful hint: Stalagmites might make it to the top.)*

## Stalagmites and Stalagtites

*Warning: NEVER go into a cave alone and always make sure someone knows where you are!*

# Caves

## Questions: Caves

1. What natural, underground hole is completely dark?
2. What kind of rock structures grow from cave floors?
3. What kind of rock structures grow from cave ceilings?

# Brown Bears

A grizzly bear is another name for a brown bear. These bears are classified as mammals because they have hair, a vertebrate, and warm blood. They also produce milk and give birth to live young.

Brown bears are omnivores. They eat berries, fruits, and nuts, as well as amphibians, fish, insects, other mammals, and small reptiles.

Brown bears live in the forests and mountains. Cougars, humans, and wolves are their main predators. The claws, size, strength, and teeth of brown bears intimidate most creatures who would try to harm them.

# Brown Bears

*Mother bears hibernate in caves for the winter to escape cold and shortage of food. In the spring, their cubs are born.*

## Questions: Brown Bears

1. What are baby bears called?
2. What is another name for brown bears?
3. Are brown bears vertebrates or invertebrates?
4. Are brown bears warm-blooded or cold-blooded?
5. Are brown bears carnivores, herbivores, or omnivores?
6. What do brown bears eat?
7. Where do brown bears live?
8. What do brown bears do in the winter to escape cold and a shortage of food?
9. What happens to female brown bears after a winter of hibernation.
10. What are the predators of brown bears?
11. What protection did God give brown bears?

# Brown Bear

Draw the other half of the brown bear in the squares below.

# Bats

Bats live in large colonies of 100 to 1,000 bats. This is one of the ways they are kept safe from hawks, owls, raccoons, and snakes. They locate food by bouncing sound off insects. This is called echolocation. Bats are considered carnivores or herbivores. Some bats eat insects; some bats eat fruit. One kind of bat even eats blood. Baby bats are called pups.

Bats are the only mammals who are warm-blooded when awake and cold-blooded when asleep. They are also the only mammals that can fly.

# Bats

*Bats live in caves, trees, bridges, mines, and other structures.*

## Questions: Bats

1. What are baby bats called?
2. Are bats vertebrates or invertebrates?
3. Are bats warm-blooded or cold-blooded?
4. Where do bats live?
5. Are bats carnivores, herbivores, or omnivores?
6. What do bats eat?
7. How do bats locate food?
8. What are the predators of bats?
9. What protection did God give bats?
10. What is the only flying mammal in the world?

# Bat Activity

Draw the other half of the bat in the squares below.
(See page 51 for complete picture.)

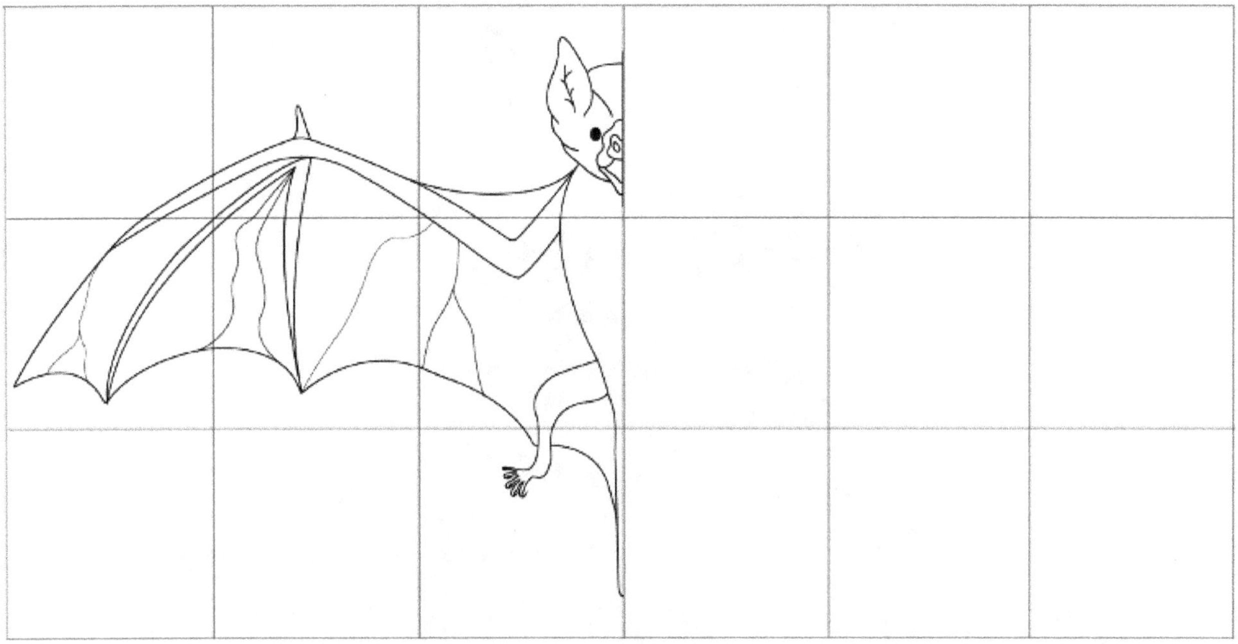

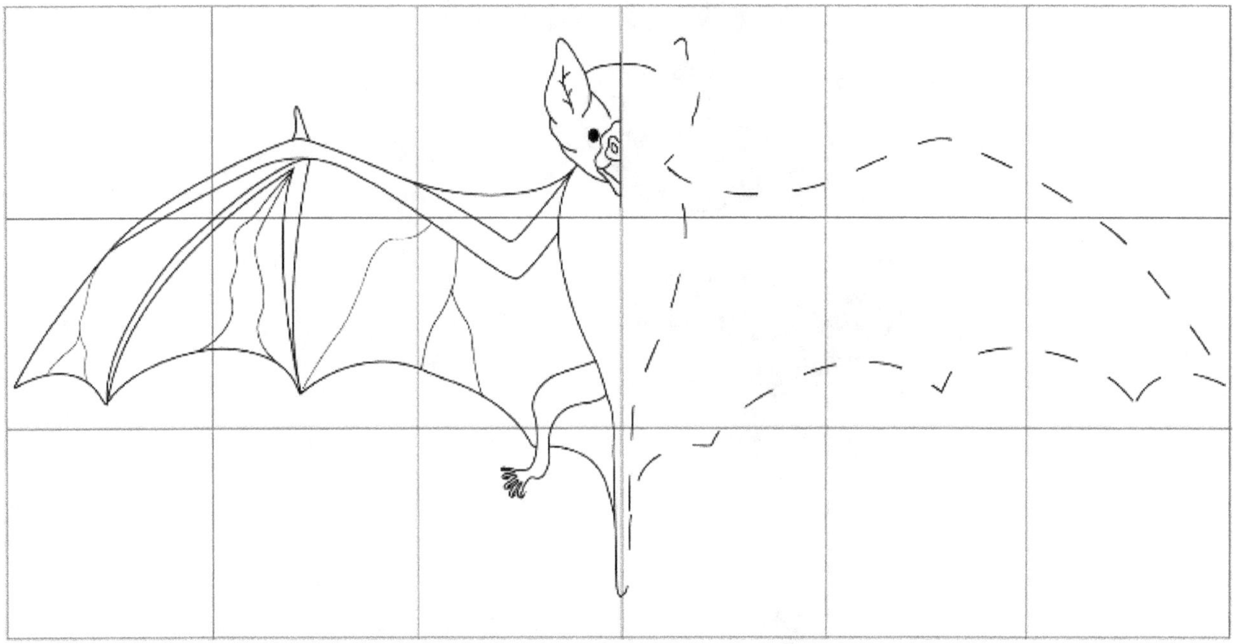

# Unit 2
## Sky and Water

*Then God said, "Let the waters teem with swarms of living creatures…" (Genesis 1:20)*

# Day 2
## Sky and Water

*God said, "Let there be an expanse in the midst of the waters, and let it separate the waters from the waters." And God made the expanse and separated the waters that were under the expanse from the waters that were above the expanse. And it was so. And God called the expanse Heaven. And there was evening and there was morning, the second day (Genesis 1:6-8)*

# Chapter 5
## Clouds

*Look at the heavens and see;*
*And look at the clouds—they are higher than you.*
*(Job 35:5)*

### Vocabulary

**Clouds:** condensed water vapor floating in the sky
**Cirrus clouds:** look like horse tails; a sign of fair weather
**Cumulus clouds:** white, puffy clouds; a sign of good weather
**Stratus clouds:** flat sheets of clouds; fog; low to the ground
**Nimbus clouds:** dark clouds; rain or snow inside them

# Clouds

Go outside and observe the types of clouds in the sky.
Color this page.

# Clouds

### Stratus

### Cirrus

### Nimbus

### Cumulus

## Questions: Clouds

1. What kind of clouds are a sign of fair weather?
2. What kind of clouds have rain or snow inside them?
3. What kind of clouds are a sign of good weather?
4. What kind of clouds look like horse tails?
5. What kind of clouds are flat sheets?
6. What kind of clouds are white and puffy?
7. What kind of clouds are foggy and low to the ground?
8. What kind of clouds are dark and stormy?

# Chapter 6
## Water Cycle

*He who calls for the waters of the sea*
*And pours them out on the face of the earth,*
*The Lord is His name.*
*(Amos 9:6b)*

### Vocabulary

**Condensation:** gas changing into liquid

**Evaporation:** liquid turning into gas

**Precipitation:** rain, snow, sleet, or hail falling to the ground

**Water Cycle:** evaporation, condensation, and precipitation

**Water Cycle**
Color this page.

# Evaporation Activity

**Purpose**
To observe evaporation in the water cycle

**Materials**
1 clear glass
Water
Permanent marker

**Procedure**
Put a glass of water on a window ledge.
Mark the water level with a permanent marker.
Check the water level after one week.

**Result**
The water will evaporate causing the water level to lower.

**Why?**
Evaporation happens when heat from the sun causes water molecules to move faster, creating water vapor.

# Condensation Activity

**Purpose**
To observe condensation in the water cycle

**Materials**
2 mason jars (heat resistent)
Electric kettle
1/2 cup hot water
4 ice cubes

**Procedure**
Boil water in kettle.
Fill one glass with hot water.
Place the other glass upside down on top of the first glass.
Put ice cubes on top of the two glasses.

**Result**
Condensation will begin to form inside the two glasses.

**Why?**
Condensation happens when air can no longer hold more water. Ice cubes speed the process.

# Precipitation
Activity

**Purpose**
To observe precipitation in the water cycle

**Materials**
1 clear 8 ounce glass
Cold water
Shaving cream
Blue food coloring

**Procedure**
Fill a glass with water.
Spray shaving cream on the top of the water to form clouds.
Saturate the shaving cream with blue food coloring.

**Result**
The "clouds" will "rain" inside the glass.

**Why?**
Water vapor gets cold and condenses into droplets of water. When the droplets become heavy enough, they fall to the earth in the form of rain, snow, sleet, or hail.

# Water Cycle

Evaporation, Condensation, Precipitation

## A Storm

### Questions: Water Cycle

1. The process of liquid turning into gas is called ________________________________.

2. The process of gas changing into liquid is called ________________________________.

3. The process of rain, snow, sleet, or hail falling to the ground is called

________________________________________.

# Chapter 7
## The Invisible

*Now unto the King eternal, immortal,* ***invisible,***
*the only wise God, be honor and glory*
*for ever and ever. Amen.*
*(I Timothy 1:17)*

### Vocabulary

**Atom:** the smallest part of any element; invisible

**Invisible:** something our eyes cannot see

**Microscope:** an instrument used to make very small objects larger

### The Invisible

God is **invisible**, but we know God exists by experiencing the effects of His creation.

Wind is **invisible**, but there is evidence of a gentle breeze or a destructive tornado.

Music is **invisible**, but voices and instruments produce beautiful melodies.

Radio waves are **invisible**, but sounds can be heard on our favorite radio stations.

Gravity and magnetism are **invisible**, but both can be scientifically proven.

The smell of lavender oil is **invisible**, but its fragrance is sweet.

Air in **invisible**, but necessary for sustaining life.

Ultraviolet rays are **invisible**, but a sunburn proves their existence.

Microscopes are used to prove the existence of things **invisible** to our eyes.

# Invisible Music
## Activity

**Purpose**
To understand "the invisible" with music

**Materials**
*Jesus Loves Me*
Guitar

**Jesus Loves Me**

```
 D
Jesus loves me this I know
  G      D
For the Bible tells me so.
  D
Little ones to Him belong.
  G        D
They are weak
     A      D
But He is strong.
 D          G
Yes, Jesus loves me.
 D            A
Yes, Jesus loves me.
 D           G
Yes, Jesus loves me.
       D   A    D
The Bible tells me so.
```

**Procedure**
Sing *Jesus Loves Me* as you play the guitar.
Turn the E string to make it out of tune.
Play and sing *Jesus Loves Me* again.

**Result**
The music will sound out of tune.

**Why?**
There is only one perfect E. If the E is either sharp or flat, it is no longer in tune. Similar to perfect pitch, the Bible provides a perfect standard (plumb line) to guide our lives. If we go off the path even a little, our lives go "out of tune".

# Invisible Ink
## Activity

*George Washington's army used invisible ink to send secret messages during the American Revolutionary War.*

**Purpose**
To observe that everything is not always how it seems

**Materials**
½ c lemon juice
Cups
3x5 cards
Small paintbrushes
Iron
Blow dryer

**Procedure**
Write a secret message or name on a card with lemon juice.
Dry the cards with a blow dryer.
With a hot iron, press the cards until words appear.
Read the words to the class.

**Result**
The words become readable.

**Why?**
Carbon compounds in lemon juice are colorless at room temperature. Any words written with lemon juice will be invisible. When heat is applied by an iron, carbon breaks down and releases carbon compounds. This is called oxidation. Oxidation can also be seen when fruit turns brown.

# Invisible Germs
## Activity

*Fingernails and chapped skin are especially hard to get clean.*

**Purpose**
To highlight the danger of germs

**Materials**
Glo Germ Gel (artificial germs)
UV flashlight

**Procedure**
Apply Glo Germ Gel to your hands.
Rub your hands back and forth until the gel is evenly distributed.
In a dark place, shine a UV light on your hands in order to see if the gel was applied evenly.
Wash your hands with soap and water.
Shine the UV light on your hands again. Notice what parts of your hands did not come clean.

**Result**
Even after washing with soap and water, the gel will cause your hands to look blue in some places when the UV light is shined upon them.

**Why?**
Doctors and nurses wash their hands thoroughly in order to prevent the spread of germs and disease. This activity is the same process used to train medical workers to wash their hands.

# The Invisible
## Activity

Draw original pictures below showing the effects of the invisible.

# Microscope
## Activity

**Purpose**
To discover the world of the invisible

**Materials**
Microscope
Hair
Onion skin
Prepared slides

**Procedure**
Observe various elements under a microscope.

**Result**
Details can be seen by a microscope far beyond what our eyes can naturally see.

**Why?**
A microscope magnifies (enlarges) objects with lenses.

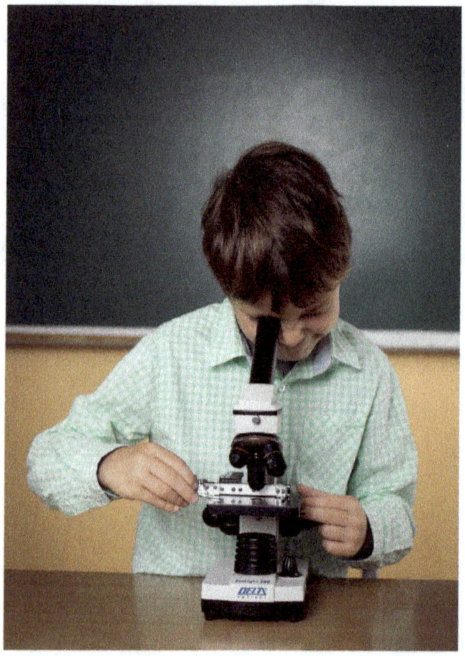

## Questions: The Invisible

1. If there is something your eyes cannot see, it is ________________.
2. The smallest part of an element is called an ________________.
3. An instrument used to make small objects larger is called a ________________.

# Chapter 8
# Float or Sink

*Immediately, Jesus made the disciples get into the boat and go on ahead of Him to the other side, while He dismissed the crowd. After He had dismissed them, He went up on a mountainside by Himself to pray.*

*Later that night, He was there alone, and the boat was already a considerable distance from land, buffeted by the waves because the wind was against it. Shortly before dawn Jesus went out to them, walking on the lake. When the disciples saw Him walking on the lake, they were terrified. "It's a ghost," they said, and cried out in fear.*

*But Jesus immediately said to them: "Take courage! It is I. Don't be afraid."*

*"Lord, if it's you," Peter replied, "Tell me to come to you on the water."*

*"Come," He said.*

*Then Peter got down out of the boat, walked on the water and came toward Jesus. But when he saw the wind, he was afraid and, beginning to sink, cried out, "Lord, save me!"*

*Immediately, Jesus reached out His hand and caught him. "You of little faith," He said, "Why did you doubt?"*

*And when they climbed into the boat, the wind died down. Then those who were in the boat worshiped Him, saying, "Truly you are the Son of God."(Matthew 14:22-33)*

# Basilisk Lizard
# "Jesus Christ Lizard"

Basilisk lizards are sometimes called Jesus Christ lizards because they can run on the water's surface at 15 mph and for 33-66 feet before sinking. Their quick speed gives them the ability to "walk on water". God gave them hinges, hollow bones, and air pockets on their feet to accomplish this feat.

Basilisk lizards are classified as reptiles. Therefore, they are cold-blooded and have a vertebrate (backbone). They lay eggs that hatch into hatchlings; they do not care for their young.

These lizards live in tropical forests. As omnivores, they eat plants, insects, smaller lizards, snails, and spiders. Birds, fish, nocturnal mammals, and snakes eat them.

## Vocabulary

**Density:** the measurement of how heavy something is compared to its size

**Float:** when an object stays on top of the water and does not sink

**Sink:** when an object goes lower than the surface of the water

## Basilisk Lizard
### "Jesus Christ Lizard"

## Questions: Basilisk Lizard

1. What is another name for a basilisk lizard?
2. How do basilisk lizards "walk on water"?
3. How fast and far do basilisk lizards run on water?
4. Are basilisk lizards mammals or reptiles?
5. Do basilisk lizards lay eggs or give birth to live young?
6. What are baby basilisk lizards called?
7. Are basilisk lizards vertebrates or invertebrates?
8. Are basilisk lizards warm-blooded or cold-blooded?
9. Where do basilisk lizards live?
10. Are basilisk lizards carnivores, herbivores, or omnivores?
11. What do basilisk lizards eat?
12. What are the predators of basilisk lizards?
13. What protection did God give basilisk lizards?

# Water Tension
## Spiritual Activity

*Behold, how good and how pleasant it is for brethren to dwell together in unity!*
*(Psalm 133:1)*

*The pepper represents people. The soap represents sin and Satan.*
*Sin blinds and causes broken people to scatter.*

**Purpose**

To make a spiritual application while learning about surface tension

**Materials**

Liquid soap
Pepper
Water
White bowl

**Procedure**

Fill a white bowl with water.
Generously sprinkle pepper on the top of the water.
Dip your finger into liquid soap.
Touch the pepper lightly with your soapy finger.

**Result**

The soap will cause the pepper
to immediately scatter to the sides of the bowl.

**Why?**

The surface tension of the water causes the pepper to float.
The soap breaks the surface tension forcing the pepper to scatter.

# Floating Egg
## Activity

**Purpose**
To observe what happens when salt, water, and an egg are mixed together

**Materials**
Glass jar
Egg
Water
Salt

**Procedure**
Fill a glass jar with water.
Place an egg in the water and watch it sink.
Add salt to the water until the egg floats to the top.

**Result**
The egg will float to the top as salt is added to the water.

**Why?**
Salt adds mass to the water making it denser. The denser the water, the easier it is for objects to float. Therefore, objects float better in salt water than in fresh water.

## Questions: Float or Sink?

1. When an object stays on top of the water and does not sink, it ________________________.
2. When an object goes lower than the top of the water, it ________________________.

## What floats? What sinks?
### Home Activity

Collect different items in your home.
Fill a container with water.
Write words or draw pictures below.

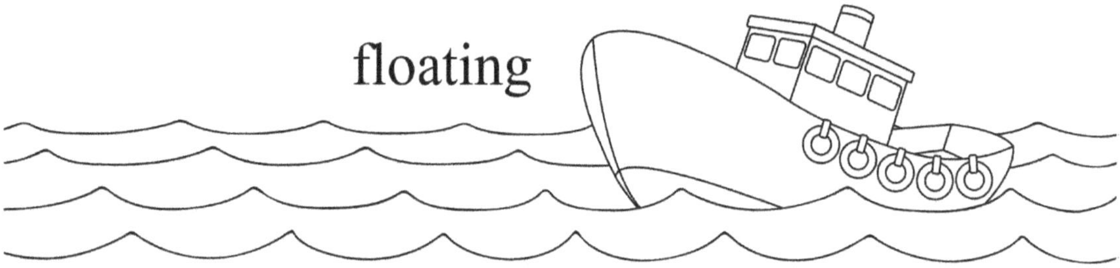

floating

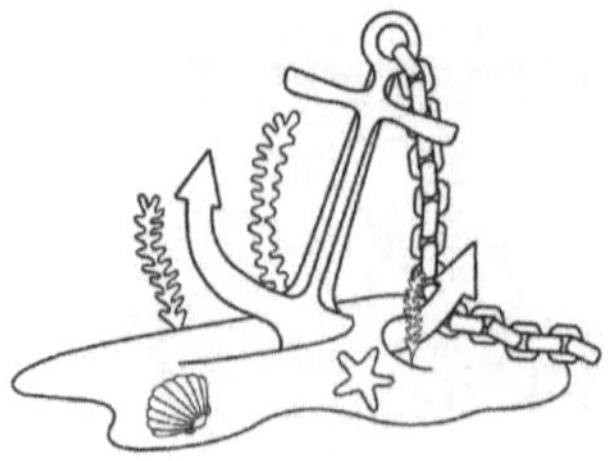

sinking

# Unit 3
## Dry Land and Plants

*But the one sown with seed on the good soil,*
*this is the one who hears the word and understands it,*
*who indeed bears fruit and produces,*
*some a hundred, some sixty, and some thirty times as much.*
*(Matthew 13:23)*

## Day 3
### Dry Land and Plants

*God said, "Let the earth sprout vegetation, plants yielding seed, and fruit trees bearing fruit in which is their seed, each according to its kind, on the earth."*
*And it was so. The earth brought forth vegetation, plants yielding seed according to their own kinds, and trees bearing fruit in which is their seed, each according to its kind. And God saw that it was good.*
*And there was evening and there was morning, the third day.*
*(Genesis 1:11-13)*

# Chapter 9
## Plants

*There is an appointed time for everything. And there is a time for every matter under heaven. A time to give birth and a time to die; A time to plant and a time to uproot what is planted.*
*(Ecclesiastes 3:1-2)*

*Truly, truly I say to you, unless a grain of wheat falls into the earth and dies, it remains alone; but if it dies, it bears much fruit.*
*(John 12:24)*

## Vocabulary

**Bud:** the earliest stage of a flower or fruit

**Flower:** the part of a plant with petals; produces fruits and seeds

**Fruit:** the edible part of a plant that contains seeds

**Leaf:** the flat, green part of a plant that grows from a stem

**Roots:** the part of a plant that grows underground to absorb water and nutrients

**Seeds:** the smallest part of a plant that grows into a new plant

**Stem:** the part of a plant that supports branches and leaves

**Transpiration:** the process of pulling water from the soil into flowers and leaves

**Xylem:** a tube in the stem that pulls water and nutrients upward from the roots

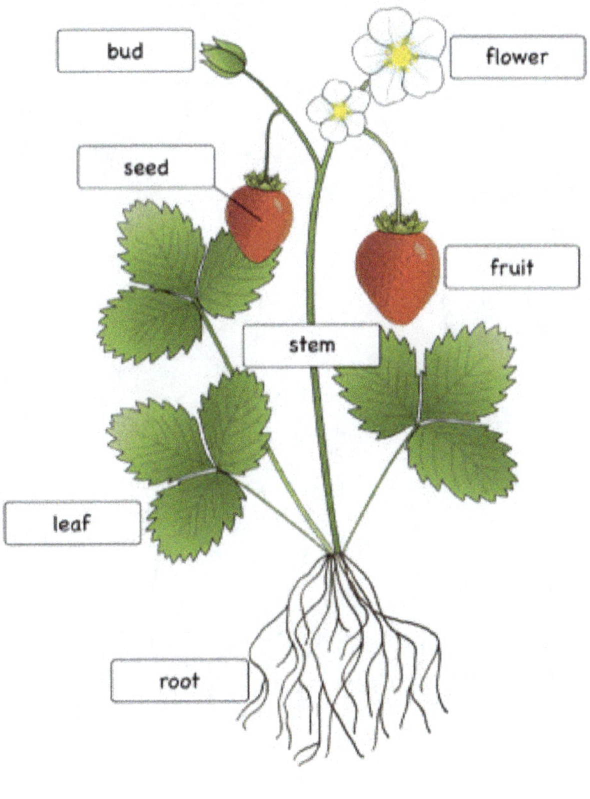

# Parts of Plants

Color and label the parts of a plant.
*(Answers on page 78)*

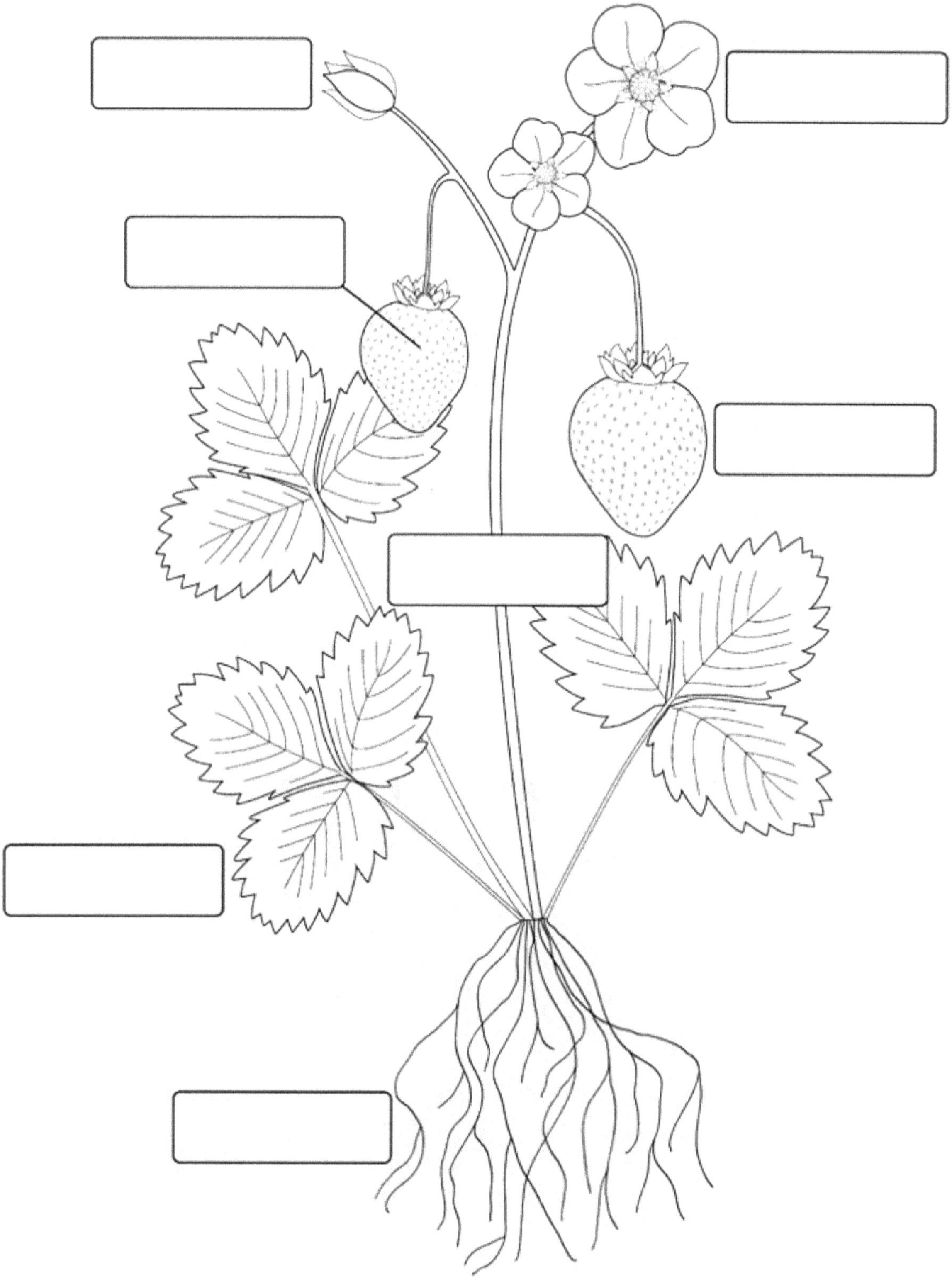

# Transpiration
## Activity

**Purpose**
To discover how water molecules are "pulled" up a stem into flowers

**Materials**
3 glass flower vases (no color or pattern)
3 white carnations
Food colors: blue, red, and yellow
Water

**Procedure**
Fill three clear vases with water.
Drip 20 drops of red food color into one of the vases.
Do the same with the yellow, and then the blue into the remaining 2 vases.
Place a white carnation into each of the vases.
Put the vases in a safe place for one week.

**Result**
The flowers will slowly change from white to the color of the water.

**Why?**
The petals change color after the xylem brings colored water through the stem.

# Plants
## Word Search

| | | | | | | | | | | | | | |
|---|---|---|---|---|---|---|---|---|---|---|---|---|---|
| D | H | T | D | S | K | V | G | R | Z | L | Y | F | R |
| A | U | V | E | E | L | W | H | S | A | M | Z | G | S |
| B | I | B | F | E | M | X | L | T | B | N | A | H | T |
| R | J | U | C | D | N | E | I | R | C | O | B | I | U |
| E | K | W | A | H | A | Y | J | O | D | P | X | J | V |
| T | I | U | R | F | O | Z | K | O | E | Q | Y | K | W |
| A | L | X | N | L | P | A | L | T | F | R | L | L | X |
| W | M | S | A | O | Q | B | M | U | G | S | E | M | Y |
| C | N | Y | T | W | R | C | N | V | T | T | M | N | N |
| D | O | Z | I | E | S | D | O | N | H | U | C | U | Z |
| E | P | A | O | R | M | E | A | W | I | V | S | O | A |
| F | Q | B | N | I | T | L | P | X | J | W | D | P | B |
| T | R | A | N | S | P | I | R | A | T | I | O | N | C |
| G | S | C | G | J | U | F | Q | Y | K | X | E | Q | D |

| | | |
|---|---|---|
| BUD | FRUIT | PLANT |
| SEED | FLOWER | CARNATION |
| LEAF | XYLEM | WATER |
| ROOT | TRANSPIRATION | SUN |

# Plants

## Questions: Plants

1. What is the earliest stage of a flower or fruit?
2. What has petals and produces fruits and seeds?
3. What part of a plant is green and flat?
4. What is very small and miraculously grows into a new plant?
5. What part of a plant supports branches and leaves?
6. What parts of some plants can you eat?
7. What tube pulls water and nutrients upward from the roots?
8. What is the process of transpiration?

# Chapter 10
## The Sower and the Seed

*[Jesus] told them many things in parables, saying, "Behold, the sower went out to sow; and as he sowed, some seeds fell beside the road, and the birds came and ate them up.*

*Others fell on the rocky places, where they did not have much soil; and they sprang up immediately, because they had no depth of soil. But after the sun rose, they were scorched; and because they had no root, they withered away. Others fell among the thorns, and the thorns came up and choked them out.*

*But others fell on the good soil and yielded a crop, some a hundred, some sixty, and some thirty times as much. The one who has ears, let him hear." (Matthew 13:3-9)*

*"When anyone hears the word of the kingdom and does not understand it, the evil one comes and snatches away what has been sown in his heart. This is the one sown with seed beside the road.*

*The one sown with seed on the rocky places, this is the one who hears the word and immediately receives it with joy; yet he has no firm root in himself, but is only temporary, and when affliction or persecution occurs because of the word, immediately he falls away.*

*And the one sown with seed among the thorns, this is the one who hears the word, and the anxiety of the world and the deceitfulness of wealth choke the word, and it becomes unfruitful.*

*But the one sown with seed on the good soil, this is the one who hears the word and understands it, who indeed bears fruit and produces, some a hundred, some sixty, and some thirty times as much." (Matthew 13:19-23)*

## Vocabulary

**Crops:** farm plants

**Plants:** living things with seeds, roots, stems, leaves, and flowers

**Soil:** the top layer of the earth

**Thorns:** sharp points located on the stems of some plants

## Farmer's Seed
### Activity

**Purpose**
To discover what plants need to grow (sun, water, and good soil)
To observe the productivity of various types of soil

**Materials**
4 plastic cups
Spade
Permanent markers
Fast-growing grass seeds
Dirt from a path
Dirt mixed with rocks
Dirt mixed with plant thorns
Good soil

**Procedure**
First cup: dirt from a path
Second cup: rocky soil
Third cup: dirt mixed thorns from bushes
Fourth cup: good soil

Label cups according to type of soil.
Pour different soils into cups as directed above.
Plant fast-growing grass seeds into all types of soils.
Water gently and place in the sun.
Daily observe the grass as it begins to grow.

**Result**
The good soil will yield the most grass. The other soils may yield a little grass, but the grass will soon wither and die.

**Why?**
Plants need sun, water, and good soil to grow.

## Seeds

## Questions: Plants

1. What are living things with seeds, roots, stems, leaves, and flowers called?
2. What are farm plants called?
3. What is the top layer of the earth called?
4. What are the sharp points on some stems called?

# Chapter 11
## Poisonous Plants

*...that there will not be among you a root bearing poisonous fruit and wormwood.*
*(Deuteronomy 29:18b)*

### Vocabulary

**Plants:** living things with seeds, roots, stems, leaves, and flowers

**Poison:** a substance found in plants that can cause illness or death

### Deadly Nightshade

*Beware, these berries can kill!*

## Water Hemlock

Water hemlock is NorthAmerica's most poisonous plant. It grows around swampy areas, Victims suffer from vomiting, unconsciousness, pains, and convulsions leading to death.

## Oleander

This flowering shrub can be found throughout the United States. Victims of this plant's poison suffer from too much saliva, diarrhea, seizures, coma, and eventually death.

## Daffodil

These flowers are white, yellow, or orange. Their poison is not strong, but if eaten, nausea, vomiting, and diarrhea will be the result.

## Deadly Nightshade

At eye level for children, these berries are sweet to the taste and look good to eat. Two to four berries will kill a child. Ten to twenty berries will kill an adult.

## Poison Oak

Poison oak is common throughout California and the northwestern states.
It can be found in just about every natural plant community where there are shrubs or trees.
The leaves contain an oily substance that causes itchy rashes and should be avoided.
It is identified by a pattern of three leaves.

## Poison Ivy

Poison ivy grows throughout the United States.
It can be found in wooded areas, along forest edges, and in urban green areas.
The leaves contain an oily substance that causes itchy rashes and should be avoided.
It is identified by a pattern of three leaves.

# Poisonous Plants

## Matching

Write the letter of the plant next to the correct photo.

1. _____ A. Poison Ivy

2. _____ B. Water Hemlock

3. _____ C. Deadly Nightshade

4. _____ D. Oleander

5. _____ E. Daffodil

6. _____ F. Poison Oak

## Mushroom

***Warning!***

*Never eat a mushroom found in your yard.*
*It could be poisonous!*

*A mushroom is technically a fungus and not a plant.*

### Questions: Poisonous Plants

1. A substance causing illness or death is _______________.
2. A poisonous plant with three leaves is called ______________ _____. (answers may vary)
3. The most poisonous plant in North America is ____________ __________________.
4. A beautiful deadly flowering shrub is called __________________.
5. A deadly plant which can kill a child with just two to four berries is called ______________ ___________________________.

## Chapter 12
## Habitats

*All things were made through Him, and without Him nothing was made that was made.*
*(John 1:3)*

## Vocabulary

**Camouflage:** the ability to blend into backgrounds by colors, patterns, or textures

**Desert:** land with very little rainfall

**Forest:** land filled with trees

**Fresh Water:** water with little or no salt

**Grassland:** land covered with different kinds of grasses

**Habitat:** any place where plants and animals live and grow

**Ocean:** salt water covering almost three-fourths of the earth's surface

**Poles:** cold land covering the top and bottom of the earth (North and South Pole)

All habitats need
AIR, SHELTER, WATER, and FOOD

## Grassland

**Animals:** bison, coyotes, deer, elk, ferrets, mountain lions, prairie dogs, giraffe, and pronghorns

**Plants:** Bermuda grass, blue fescue, elephant grass, feather grass, lemongrass, red oats grass, and Rhodes grass

## Forest

**Animals:** badgers, bears, chipmunks, deer, foxes, rabbits, raccoons, and squirrels,

**Plants:** ferns, mosses, shrubs, trees, vines, and wildflowers

## North and South Poles

**Animals:** bald eagles, deer, moose, musk ox, penguins, polar bears, reindeer, and seals

**Plants:** dwarf woody shrubs, grasses, lichens, and mosses

## Desert

**Animals:** birds, bobcats, coyotes, lizards, rattlesnakes, scorpions, and spiders

**Plants:** acacias, cacti, creosote bush, mesquite, succulents, and yucca

## Fresh Water

**Animals:** alligators, catfish, dolphins, eels, frogs, otters, turtles, and water snakes

**Plants:** algae, aspen, cattails, water lilies, and willow trees

## Ocean

**Animals:** harbor seals, lion fish, otters, sea lions, sharks, turtles, and whales

**Plants:** kelp, phytoplankton, red algae, sargassum, and seagrass

# Habitats

## Matching

Write the letter of the habitat next to the correct photo.

1.________

2.________

3.________

4.________

5.________

6.________

A. Fresh Water

B. Desert

C. Ocean

D. Grasslands

E. Forest

F. Poles

# Habitats
## Activity

**Purpose**
To learn about habitats

**Materials**
6 Hula hoops
Paper
Markers, colored pencils, or crayons
Numerous small toy animals

**Procedure**
Draw pictures of six habitats
(grassland, forest, desert, poles, salt water, and fresh water).

Place six hula hoops on the ground.
Place one habitat picture inside each hoop.
Place each animal in its correct habitat.

**Result**
Animals will be categorized into different types of habitats.

**Why?**
God placed animals in various kinds of habitats according to needs.
A habitat is any place where plants and animals live and grow.

## Farm Habitat

Write or draw different kinds of animals living on a farm.

# Habitats

## Questions: Habitats

1. What are four things you need in a habitat to survive?
2. How many days can you live without water?
3. What habitat has a lot of trees?
4. What habitat has salt water?
5. What habitat is very cold?
6. What habitat contains different types of grasses?
7. What habitat has a lot of water with little or no salt?
8. What habitat does not have much rainfall?
9. What habitat do you live in?

# Unit 4
## Sun, Moon, Stars, and Planets

## Day 4
### Sun, Moon, Stars, and Planets

*14 And God said, "Let there be lights in the expanse of the heavens to separate the day from the*
*night. And let them be for signs and for seasons,[a] and for days and years, 15 and let them be lights*
*in the expanse of the heavens to give light upon the earth." And it was so. 16 And God made the two*
*great lights—the greater light to rule the day and the lesser light to rule the night—and the stars. 17*
*And God set them in the expanse of the heavens to give light on the earth, 18 to rule over the day and*
*over the night, and to separate the light from the darkness. And God saw that it was good. 19 And*
*there was evening and there was morning, the fourth day.*
*(Genesis 1:14-19)*

# Chapter 13
## Sun and Other Stars

*Thus says the LORD: Who gives the sun for a light by day, The ordinances of the moon and the stars for a light by night…The LORD of hosts is His name.*
*(Jeremiah 31:35)*

*There will be no more night. They will not need the light of a lamp or the light of the sun, for the Lord God will give them light. And they will reign forever and ever.*
*(Revelation 22:5)*

## Vocabulary

**Astrology:** using constellations to predict events

**Astronomy:** scientifically-based study of the universe

**Constellation:** a group of stars identified according to pattern

**Orbit:** a repeating path one object takes around another object

**Shadows:** dark images created when objects block light

**Stars:** lights in the night sky

**Sun:** a star that gives the earth heat and light

*The sun provides the earth's heat and light. All eight planets orbit around it. Even though the sun looks orange, it really is white.*

# Constellations

*He counts the number of the stars;*
*He calls them all by name.*
*(Psalm 147:4)*

*He made the Bear, Orion, and the Pleiades,*
*And the chambers of the south.*
*(Job 9:9)*

*Can you bring out Mazzaroth in its season?*
*Or can you guide the Great Bear with its cubs?*
*(Job 38:32)*

***Mazzaroth:*** *Hebrew word for "constellation"*

**Ursa Minor/Cub** (includes Little Dipper)
**Ursa Major/Mother Bear** (includes Big Dipper)

"Astrology is based on understanding the positions of the stars, which seems like a scientific enough pursuit in itself. But is there any science to back up whether astrology impacts our personality and our lives? Here's the short answer: **No. None whatsoever."**
Scientific America, June 2020

# Constellations

There are eighty-eight constellations recognized by the International Astronomical Union (IAU). Some farmers have used constellations to determine when to plant and harvest. Captains of ships have used the stars as a navigational tool by using the two stars at the end of the Big Dipper which point to the North Star.

## Orion

A giant hunter going out to battle with a shield, a belt, and a sword

## Pleiades

A smaller version of the Big Dipper

# My Shadow

by

Robert Louis Stevenson

I have a little shadow that goes in and out with me,
And what can be the use of him is more than I can see.
He is very, very like me from the heels up to the head;
And I see him jump before me, when I jump into my bed.
The funniest thing about him is the way he likes to grow—
Not at all like proper children, which is always very slow;
For he sometimes shoots up taller like an India-rubber ball,
And he sometimes gets so little that there's none of him at all.
He hasn't got a notion of how children ought to play,
And can only make a fool of me in every sort of way.
He stays so close beside me; he's a coward you can see;
I'd think shame to stick to Nursie as that shadow sticks to me!
One morning, very early, before the sun was up,
I rose and found the shining dew on every buttercup;
But my lazy little shadow, like an arrant sleepy-head,
Had stayed at home behind me and was fast asleep in bed.

## Shadows
### Activity

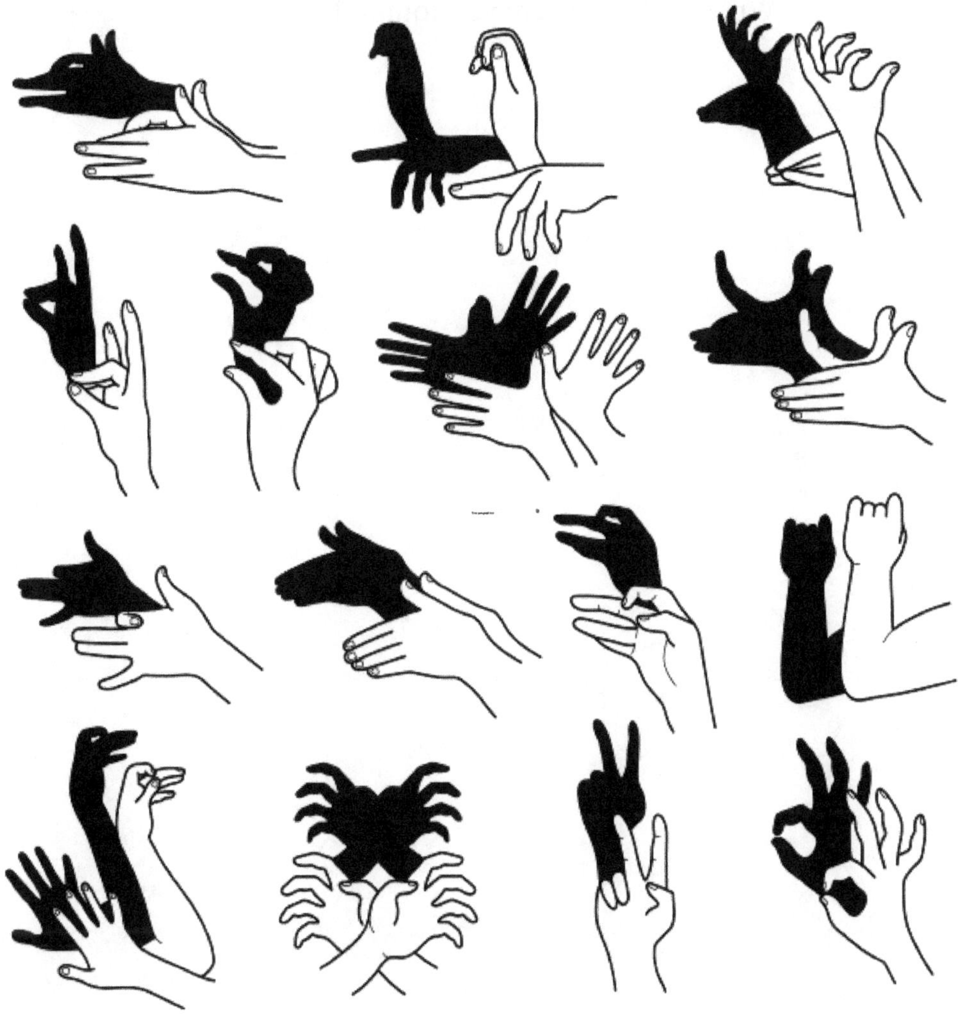

**Purpose**
To create shadows

**Materials**
Light source (lamp or projector)
Hand shadows (below)

**Procedure**
Shine a light on a wall.
Form the shapes of animals by blocking the light with your hands

**Result**
Animal shadows will form on the wall.

**Why?**
Shadows are created by blocking rays of light.

# Shadows
## Activity

Find ten shadows around your home or school.
Write or draw where you found them below.

1.

2.

3.

4.

5.

6.

7.

8.

9.

10.

## Shadows

### Questions: Outer Space

1. What star provides the earth's heat and light?
2. Our universe's eight planets orbit around which star?
3. What is the true color of the sun?
4. What are shadows?
5. What is the difference between astronomy and astrology?
6. What lights can be seen in the night sky?
7. How did farmers use constellations?
8. How many different constellations are there?
9. For centuries, how did seafarers use the Big Dipper?

# Chapter 14
## Earth's Moon

*God made the two great lights, the greater light to govern the day, and the lesser light to govern the night; He made the stars also. (Genesis 1:16-18)*

### Vocabulary

**Armstrong, Neil:** the first man to walk on the moon

**First Quarter:** a phase of the moon; looks like a D

**Full Moon:** a phase of the moon; looks like a white circle

**Last Quarter:** a phase of the moon; looks like a C

**Moon:** a planet's satellite

**New Moon:** a phase of the moon when it cannot be seen

**Satellite:** a moon, planet, or machine orbiting a planet or a star

## Neil Armstrong

*That's one small step for man, one giant leap for mankind.*
*(Neil Armstrong)*

*First man to walk on the moon on July 21st, 1969*

## Phases of the Moon

Lunar phases are caused by the position of the moon in relation to the sun.
The moon would be completely dark
without the sun's light reflecting off of it.

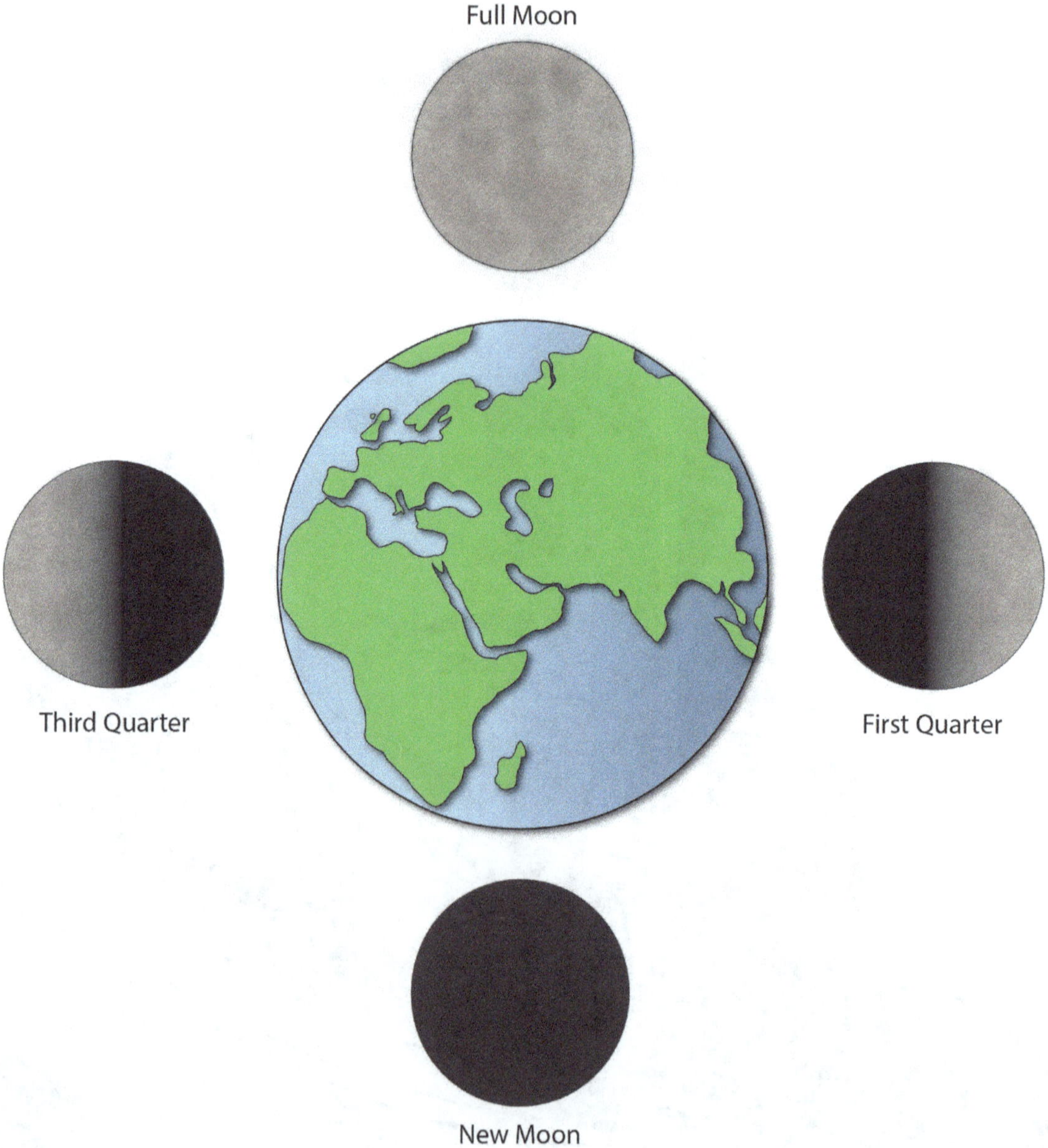


*As we are imitators of God, the Son's (Jesus') light reflects off of us.*

# Phases of the Moon
## Activity

It takes twenty-eight days for the moon
to rotate around the earth.
During the night, phases of the moon can be seen
due to this rotation and the sun's reflection.

**Purpose**
To understand phases of the moon

**Materials**
Paper plates
Phases of the Moon (page 110)
Chocolate sandwich cookies with white filling (4 per student)
White glue
Scissors

**Procedure**
Cut out the phases of the moon.
Glue the circle onto a paper plate.
Create each moon phase with chocolate sandwich cookies. (Directions below)

**Full Moon**
Create a white circle by taking the chocolate top off of a cookie.
Glue it on the circle labeled Full Moon.

**First Quarter**
Take the top off of another cookie.
Scrape off half the white frosting creating the letter D.
Glue it onto the circle labeled First Quarter.

**New Moon**
Take the cookie without any frosting.
Glue it on the circle labeled New Moon.

**Third Quarter**
Take the top off of another cookie.
Scrape off half the white frosting creating the letter C.
Glue it on the circle labeled Third Quarter.

**Result**
A visual representation of the main four lunar phases will be created.

**Why?**
The moon phases are determined by the position of the moon in relation to the sun.

## Phases of the Moon
### Activity

Go outside at night and look at the moon.
Circle the lunar phase you observe.

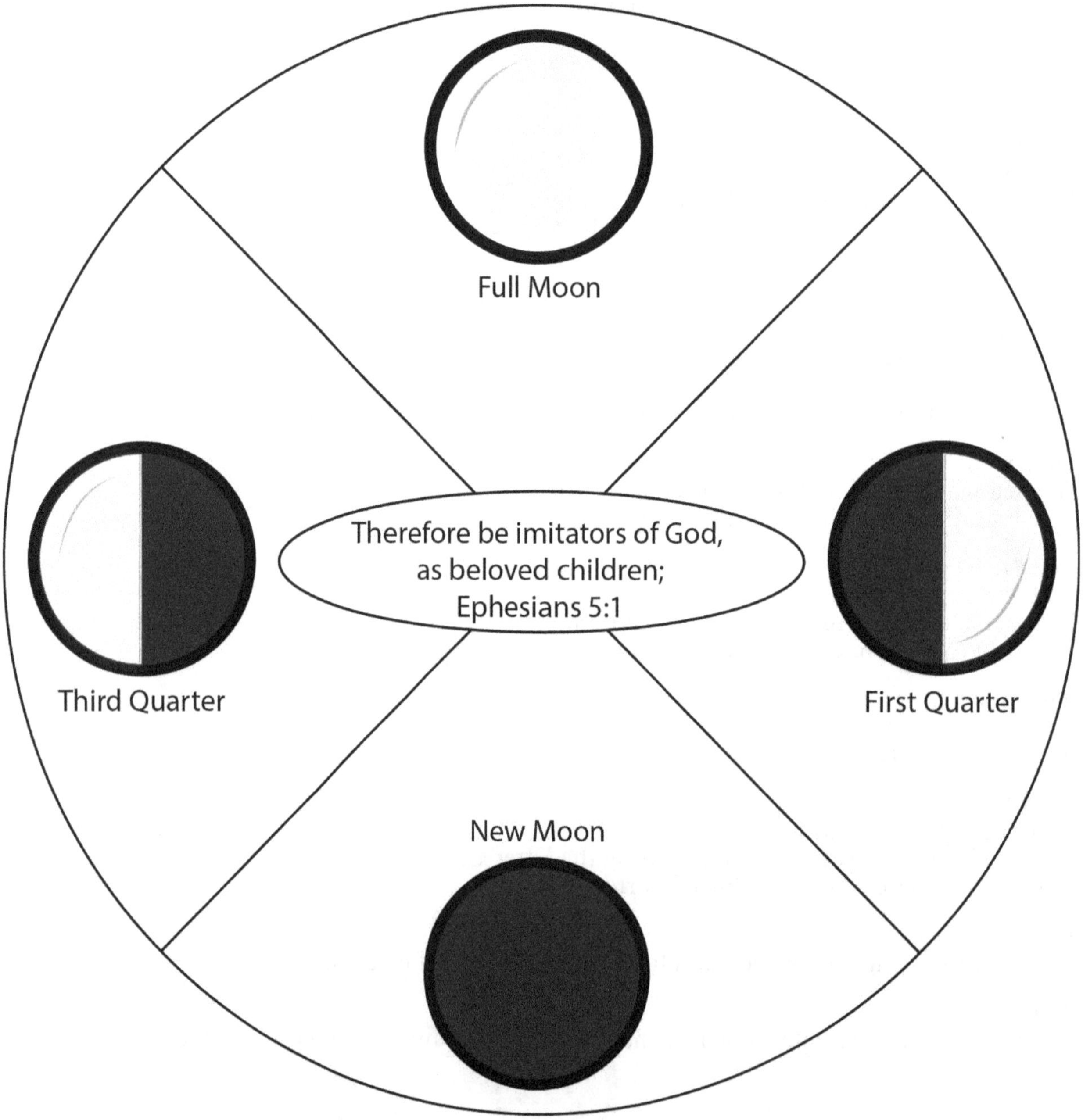


**Phase 1:** New Moon-cannot be seen

**Phase 2:** First Quarter-looks like a D

**Phase 3:** Full Moon-round white circle

**Phase 4:** Third Quarter-looks like a C

## Full Moon

### Questions: Moon

1. How many days does it take for the moon to rotate around the earth?
2. What is a satellite?
3. Does the moon reflect or refract light from the sun?
4. Who was the first man to walk on the moon?
5. What phase of the moon looks like a D?
6. What phase of the moon looks like a white circle?
7. What phase of the moon looks like a C?
8. What phase of the moon cannot be seen?

# Chapter 15
## Planets

*By faith we understand that the world has been created by the word of God so that what is seen has not been made out of things that are visible.*
*(Hebrews 11:3)*

## Vocabulary

**Planets:** large masses that orbit the sun

**Rings:** the circles around some planets made of rock, ice, and dust particles

**Solar System:** the eight planets and their moons in orbit around the sun

**Telescope:** an instrument used to collect and magnify the light from distant objects

### Telescope

# Planets

**Mercury**
Roman god of commerce, travel, and thievery
No moon or rings
Moves quickly in the sky

**Venus**
Roman goddess of love and beauty
No moon or rings
Second brightest natural object in the sky

**Earth**
English/German word meaning "ground"
One moon and no rings
Only planet with life

**Mars**
Roman god of war
Two moons and no rings
Called the "Red Planet" because of the high amount of iron oxide

**Jupiter**
King of Roman gods
Ninety-two known moons (2023) and several faint rings made of dust
Largest planet

**Saturn**
Roman god of agriculture and wealth
Eighty-three moons (2023) and bright rings made of ice and rock
Second-largest planet

**Uranus**
Greek god of the sky
Twenty-seven small moons and thirteen faint rings
Icy and burning hot with extreme seasons

**Neptune**
Roman god of the sea
At least fourteen known moons and six known narrow rings
Gravity similar to the earth's gravity

## Solar System
### Word Search

| | | | | | | | | | | | | | |
|---|---|---|---|---|---|---|---|---|---|---|---|---|---|
| A | P | B | O | N | M | J | C | A | G | D | I | E | S |
| B | K | C | S | B | I | A | S | Y | J | Z | G | H | O |
| A | J | R | L | K | H | W | X | U | H | G | R | F | L |
| Z | A | D | I | U | S | T | P | S | N | P | F | O | A |
| M | Y | V | J | T | R | I | Q | U | F | A | Q | N | R |
| X | M | W | O | N | T | A | E | V | A | O | R | L | S |
| U | I | E | U | E | M | U | J | P | C | N | K | U | Y |
| H | S | O | R | G | E | F | E | I | N | E | U | M | S |
| U | E | P | V | C | L | H | D | C | R | P | B | S | T |
| A | V | W | I | T | U | W | T | O | U | T | E | A | E |
| Y | X | E | S | Y | X | R | Z | R | T | U | L | K | M |
| T | R | Q | P | D | I | O | Y | N | A | N | M | O | A |
| S | U | N | E | V | N | M | L | G | S | E | O | I | J |
| R | Z | Q | U | H | S | T | A | R | B | N | K | A | C |

| | | |
|---|---|---|
| MARS | JUPITER | EARTH |
| VENUS | URANUS | NEPTUNE |
| MERCURY | SATURN | SUN |
| MOON | SOLAR SYSTEM | STAR |

# Planets

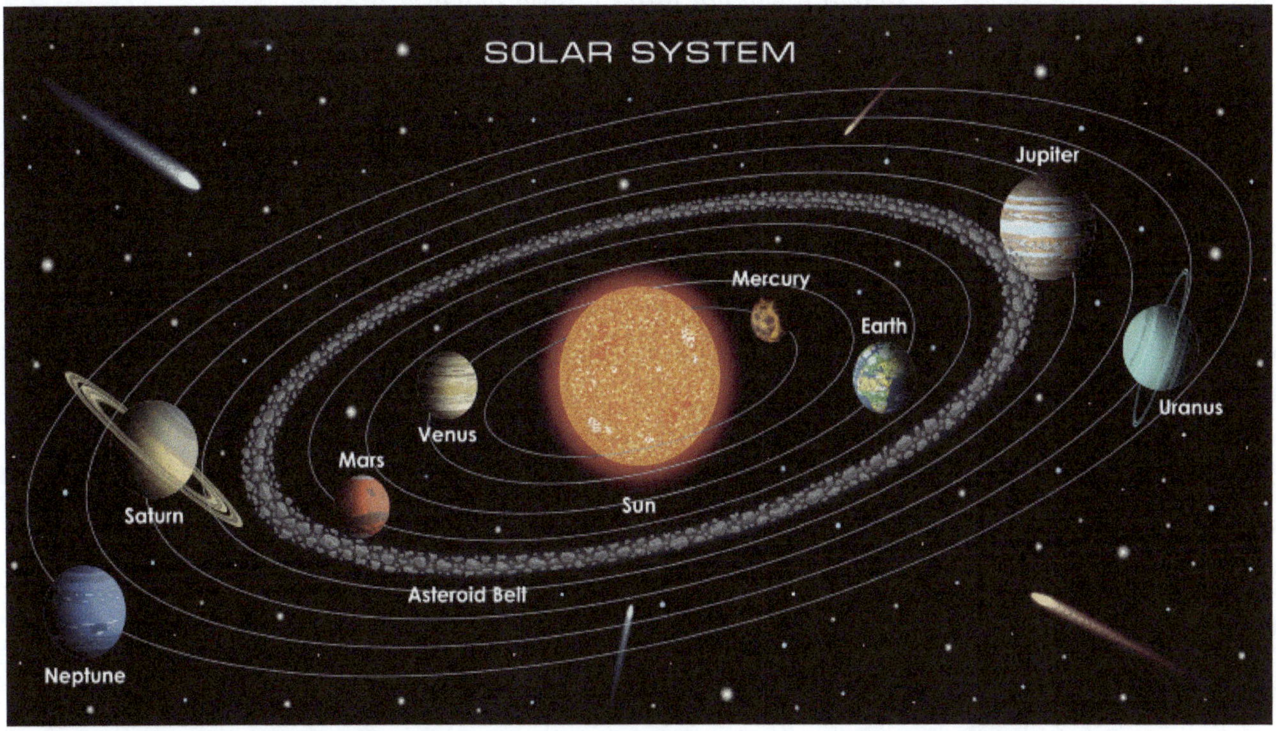


## Questions: Planets

1. What planet is closest to the sun?
2. What planet is farthest from the sun?
3. What planet has bright rings around it?
4. What planet has plants and animals on it?
5. What is the largest planet in our solar system?
6. What is the red planet?
7. How many planets are in our solar system?
8. Are any of these planets bigger than the sun?
9. What is a telescope?

# Chapter 16
## Gravity and Magnetism

*He established the earth upon its foundations,*
*So that it will not totter forever and ever.*
*(Psalm 104:5)*

*He is before all things, and in Him all things hold together.*
*(Colossians 1:17)*

## Vocabulary

**Acceleration:** to increase rate or speed

**Attract:** to pull towards

**Compass:** an instrument with a magnetized pointer indicating north

**Force:** the push or pull of an object

**Galilei, Galileo:** a famous scientist who studied gravity

**Gravity:** an invisible force that pulls objects towards the center of the earth

**Magnetism:** an invisible magnetic force observed when magnets attract or repel

**Mass:** the amount of matter in an object

**Newton, Isaac:** the first person to define a theory of gravity

**Repel:** to push away

# Gravity
## Activity

**Purpose**
To observe two objects with different weights falling to the ground

**Materials**
Basketball
Tennis ball

**Procedure**
Stand on a chair.
Hold the basketball and tennis ball even with each other.
Drop the two balls at exactly the same time.

**Result**
Even though the force of gravity pulls stronger on heavier objects, heavier objects require more force to accelerate. Thus, the two balls land on the ground at exactly the same time.

**Why?**
No matter how different the mass is between two objects, the acceleration due to gravity remains the same.

## Galileo Galilei

*Between 1589 and 1592, Galileo Galilei dropped two objects from the Tower of Pisa. He noticed they dropped to the ground at exactly the same rate, regardless of weight or size.*

# Magnetism

**Purpose**
To learn how to make a temporary magnet

**Materials**
Magnets of varying kinds and sizes
Paperclips

**Procedure**
Rub a paperclip on a magnet fifty times in the same direction.
Put it near another metal (iron).

**Result**
The paperclip will turn into a temporary magnet and attract other magnets.
If it doesn't work, rub the paperclip fifty more times and try again.

**Why?**
Temporary magnets can be created by exposing metals with iron to magnetic field.

# Isaac Newton

*Isaac Newton sat under a tree thinking when an apple fell on his head.*
*"Why did the apple fall down, and not up or sideways?" he wondered.*
*He decided, "What goes up must come down,"*
*resulting in today's concept of gravity.*

## Questions: Gravity and Magnetism

1. What is acceleration?
2. Who dropped two objects from the Tower of Pisa?
3. Who is credited for discovering gravity?
4. What is mass?
5. What is the push or pull of an object?
6. What force pulls objects to the center of the earth?
7. What happens if you drop two objects of differing weights from the same height?
8. What is the difference between attract and repel?

# Unit 5
## Fish, Boats, Birds, and Airplanes

# Day 5
## Fish, Boats, Birds, and Airplanes

*God said, "Let the waters swarm with swarms of living creatures, and let birds fly above the earth across the expanse of the heavens." So, God created the great sea creatures and every living creature that moves, with which the waters swarm, according to their kinds, and every winged bird according to its kind. And God saw that it was good. And God blessed them, saying, "Be fruitful and multiply and fill the waters in the seas, and let birds multiply on the earth." And there was evening and there was morning, the fifth day. (Genesis 1:20-23)*

# Chapter 17
## Fish

*The birds of the sky, and the fish of the sea,*
*Whatever passes through the paths of the seas.*
*(Psalm 8:8)*

## Vocabulary

**Camouflage:** the ability to blend into backgrounds by colors, patterns, or textures

**Cold-blooded:** an animal that takes its body temperature from the environment

**Fins:** the part of a fish used for movement and steering

**Fish:** cold-blooded water vertebrates who have gills, fins, and lay eggs

**Gills:** the part of the body that fish (and amphibians) use to breathe oxygen from water

**Invertebrate:** an animal without a backbone

**Predator:** an animal hunting another animal

**Prey:** an animal hunted by another animal

**Scales:** rigid plates that grow on the skin of a fish

**Tail:** the part at the end of a fish used for swimming and steering

**Warm-blooded:** an animal that maintains a constant body temperature

**Vertebrate:** an animal with a backbone

## Fish

There are over 3,000 species of fish. Fish are all cold-blooded vertebrates that rest but never sleep. Fish eat algae, amphibians, crustaceans, fish eggs, insects, mollusks, plankton, plants, and worms. Whales and dolphins look like fish, but they are actually mammals.

# Fish

## Exception to the Rule

Some fish give birth to baby fish instead of laying eggs.

**Blue Shark**

**Swordtail**

**Guppy**

**Platy**

**Molly**

## Questions: Fish

1. How many different species of fish are there?
2. Are fish vertebrates or invertebrates?
3. Are fish warm-blooded or cold-blooded?
4. Do MOST fish birth live babies or lay eggs?
5. What do fish eat?
6. Do fish sleep?
7. Are whales and dolphins considered fish?

# Sharks

Sharks are saltwater fish at the top of the food chain with no natural predators. Sharks are carnivores that survive by eating other fish. However, humans hunt sharks to use them for food, medicines, and vitamins.

Sharks have great eyesight, swim fast, and have many sets of replaceable teeth. They also have great night vision and can even see colors.

**Whale Shark**

**Great White Shark**

**Hammerhead Shark**

In the same way no human fingerprint is the same, every spot pattern is unique to each whale shark.

## Questions: Sharks

1. How do humans benefit from sharks?

2. What is the habitat of a shark?

3. What is unique to a whale shark?

4. Is a shark a carnivore, herbivore, or omnivore?

5. What do sharks eat?

6. What kind of eyesight do sharks have?

7. What are the predators of a shark?

8. What protection did God give sharks?

## Bluefin Tuna

Like the shark, bluefin tuna live in saltwater and are highly-prized for their meat. Their predators include billfishes, toothed whales, sharks, and humans. Bluefin tuna eat young bluefin and tiny zooplankton. These fish can live up to thirty years or more. Female bluefin lay up to 10 million eggs per year.

## Questions: Bluefin Tuna

1. What is a benefit of bluefin tuna fish?
2. What is the habitat of bluefin tuna fish?
3. What do bluefin tuna fish eat?
4. What are the predators of bluefin tuna fish?
5. What is the life span of a bluefin tuna?
6. How many eggs does a female bluefin tuna produce per year?

# Atlantic Salmon

*Their scales have a silver sheen with pockets of blue, crimson, purple, pink, or green most of the year. Their scales change to a rainbow of colors when reproducing.*

Atlantic salmon are born in freshwater, migrate to a saltwater sea, and return to freshwater to reproduce. The female salmon can lay up to 4,000 eggs at a time.

They eat crayfish, eels, grasshoppers, herring, krill, mayflies, shrimp, squid, stoneflies, and zooplankton. Bears, eagles, humans, otters, seals, and sharks eat them. They weigh anywhere from 3-126 pounds.

## Questions: Atlantic Salmon

1. Are salmon freshwater or saltwater fish?
2. What color are salmon scales?
3. How many eggs can a female salmon lay?
4. How big can a salmon get?
5. What do salmon eat?
6. What are the predators of salmon?

# FISH
## WORD SEARCH

| | | | | | | | | | | | | | |
|---|---|---|---|---|---|---|---|---|---|---|---|---|---|
| A | D | I | D | H | Q | A | E | F | T | B | E | H | S |
| T | E | H | J | I | R | B | I | M | A | C | G | I | C |
| B | U | T | K | E | S | S | F | N | Z | D | L | M | A |
| C | F | N | E | C | H | A | B | I | T | A | T | P | L |
| G | R | U | A | U | T | L | G | O | Y | F | Q | U | E |
| Q | O | V | F | D | U | T | H | P | T | S | T | Z | S |
| M | T | L | G | O | V | W | I | Q | U | Y | B | C | D |
| N | A | W | K | R | S | A | L | M | O | N | E | F | G |
| O | D | X | L | P | W | T | J | U | R | X | J | I | H |
| P | E | Y | M | E | X | E | K | V | T | W | K | L | M |
| S | R | Z | N | R | Y | R | L | S | X | V | N | O | P |
| G | P | A | O | J | A | C | M | I | G | R | A | T | E |
| G | R | B | P | H | Z | D | K | R | W | U | V | W | Y |
| E | S | C | S | R | E | T | A | W | H | S | E | R | F |

| | | |
|---|---|---|
| FISH | HABITAT | SALTWATER |
| SHARK | SALMON | TROUT |
| TUNA | EGGS | FRESHWATER |
| REPRODUCE | MIGRATE | PREDATOR |

# Chapter 18
## Boats

*Such large crowds gathered around [Jesus] that He got into a boat and sat in it, while all the people stood on the shore.*
*(Matthew 13:2)*

## Vocabulary

**Bow:** the front of a boat

**Fins:** the part of a fish used for movement and steering

**Gills:** the part of fish (and amphibians) used to breathe oxygen from water

**Head:** the front of a fish

**Hull:** the body of a boat

**Keel:** lower structure of a boat from bow to stern

**Mast:** a long pole that rises from the bottom of a ship and supports the sails and rigging

**Rudder:** the part used by boats for steering

**Sail:** the piece of fabric used to move a boat forward by wind

**Scales:** bony plates protecting the bodies of fish (and reptiles)

**Stern:** the back of a boat

**Tail:** the part at the end of a fish used for swimming and steering

**Trunk:** the middle part of a fish

# Boat and Fish Compared

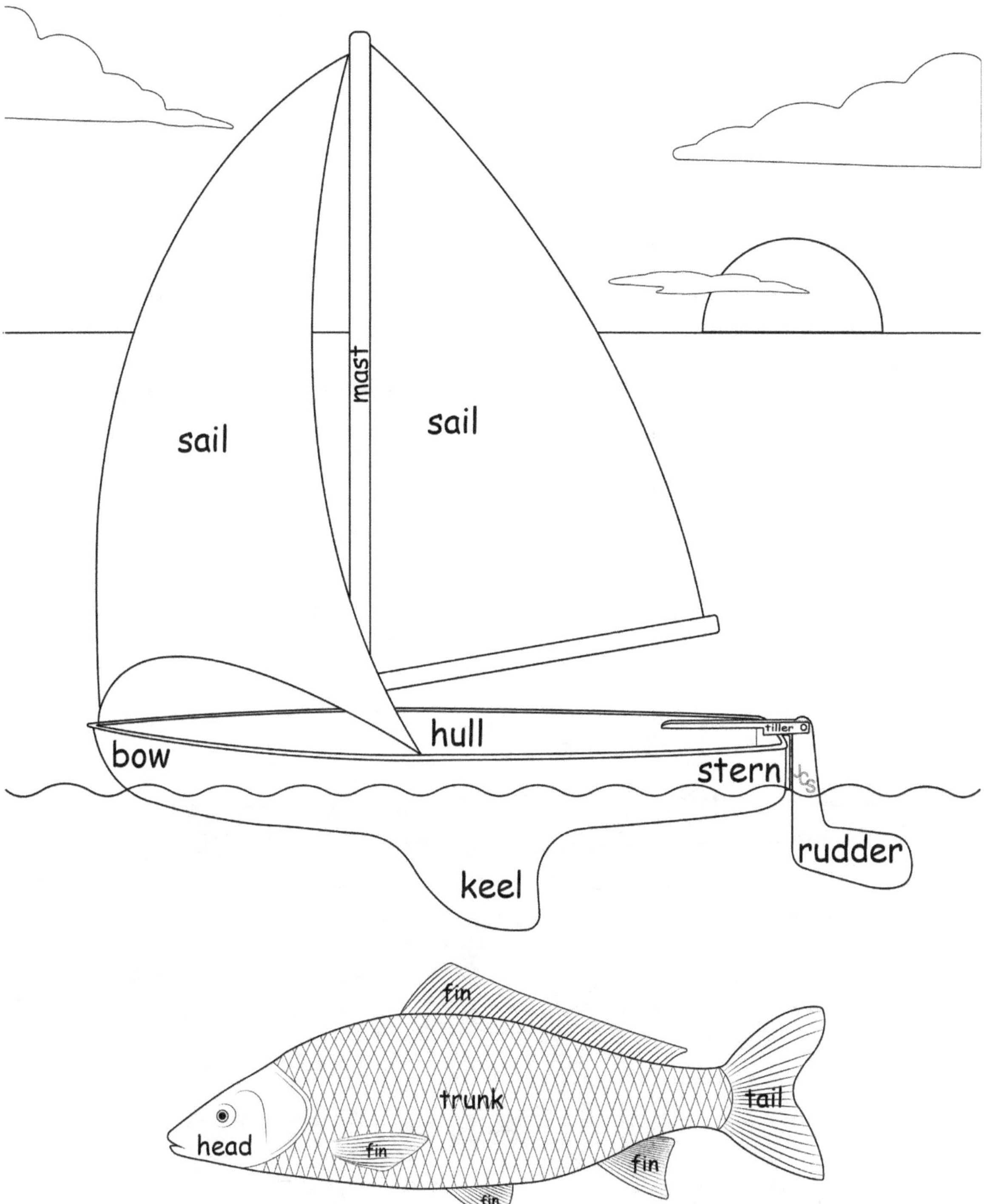

## Boat and Fish Compared

Label and color the parts of a boat and a fish.

*(Answers on page 132)*

## Sailboat

## Questions: Boats and Fish

**1.** What is the back of a boat called? What is the end of a fish called?

2. What is the middle of a boat called? What is the middle of a fish called?

3. What is the body of a boat called?

4. What covers the body of a fish?

5. What helps a boat steer? What helps a fish steer?

6. What is the front of a boat called? What is the front of a fish called?

7. What helps to move a boat forward with wind?

8. What body part is used by fish to breathe oxygen from water?

# Chapter 19
## Birds

*Look at the birds of the sky, that they do not sow, nor reap, nor gather crops into barns, and yet your heavenly Father feeds them. Are you not much more important than they?*
*(Matthew 6:26)*

## Vocabulary

**Beak:** the mouth of a bird

**Carnivore:** a meat-eating animal

**Feathers:** the covering of a bird's body

**Herbivore:** a plant-eating animal

**Omnivore:** a plant and meat-eating animal

**Vertebrate:** an animal with a backbone

**Wing:** the part of a bird used for flying

## Birds

ALL birds are vertebrates, lay eggs, are warm blooded, and have feathers. Depending on the species, they can be carnivores, herbivores, or omnivores. According to the International Ornithological Committee, there are an estimated 11,000 different species of birds in the world.

# Birdhouse
## Activity

*'My heart is like a singing bird.*
*Christina Rossetti*

**Purpose**
To observe and identify local birds

**Materials**
Milk cartons
Scissors
Paint
Buttons
Stickers
Twine
Peanut butter
Birdseed
Dowel

**Procedure**
Cut a door into a milk carton.
Put a perching dowel below the door.
Decorate.
Put peanut butter and birdseed inside the carton.
Hang with twine.

**Result**
Local birds and squirrels will visit the birdhouse for free food.

**Why?**
Bird-feeders allow for easy observation and identification of local birds.

# Birds

## Exception to the Rule

Some birds have wings but do not fly.
There are approximately 57 flightless bird species. Here are a few:

**Ostrich**

**Penguin**

**Kiwi**

**Emu**

## Questions: Birds

1. Do all birds give birth to live young, or do they lay eggs?
2. Are birds warm-blooded or cold-blooded?
3. Are all birds vertebrates or invertebrates?
4. What is a bird's mouth called?
5. What covers most of a bird's body?
6. Do MOST birds fly?

# Eagles

*Yet those who wait for the Lord*
*Will gain new strength;*
*They will mount up with wings like eagles,*
*They will run and not get tired,*
*They will walk and not become weary.*
*(Isaiah 40:31)*

Eagles are at the top of the food chain with no predators except humans. These birds are bold, cunning, and intelligent. Their eyes are eight times stronger than human eyes.

Eagles typically mate for life. If one dies, an eagle might find a new mate living in the same territory. Eagles build their nests on high cliffs and trees where the females take care of the eaglets. Males find the food consisting of amphibians, crabs, fish, and reptiles.

## Questions: Eagles

1. What are baby eagles called?

2. What do eagles eat?

3. Are eagles carnivores, herbivores, or omnivores?

4. What are the predators of an eagle?

5. Do eagles mate for life?

6. Does the male or female eagle take care of their young?

7. Where do eagles build nests?

8. What is the difference between eagle eyes and human eyes?

9. What strengths do eagles have?

## Vultures

*These are the birds you are to regard as unclean and not eat because they are unclean: the eagle, the vulture…*
*(Leviticus 11:13a)*

Vultures are carnivores that eat dead things. They prevent disease by devouring dead animals. Vultures spend a lot of time soaring and can smell a decaying animal from a mile away. They have very few predators. Although, if given a chance, hawks or eagles may eat baby vultures called chicks.

Vultures mate for life. They do not build nests, but lay eggs in dense thickets, hollow logs, and caves. Both male and female vultures take care of their chicks for about three months. Vultures form bonds with each other that last their whole lives (about 10 years). Vultures spend time together all year, not just during the breeding season.

## Questions: Vultures

1. What are baby vultures called?
2. Are vultures carnivores, herbivores, or omnivores?
3. What do vultures eat?
4. What are the predators of vultures?
5. Does the male or female vulture take care of their young?
6. Do vultures spread disease?
7. How good is the vulture's sense of smell?
8. What do vultures spend a lot of time doing?
9. Where do vultures build their nests?
10. Do vultures mate for life?

## Ravens

*Then it came about at the end of forty days, that Noah opened the window of the ark which he had made; and he sent out a raven, and it flew here and there until the water was dried up from the earth.*
*(Genesis 8:6-7)*

Ravens are omnivores, eating dead animals, eggs, grains, insects, rodents, and worms. Goshawks, hawks, and owls eat them. Ravens perform elaborate chases, dives, and rolls during mating season; they mate for life. Both parents take care of their chicks.

## Questions: Ravens

1. What is a baby raven called?
2. Are ravens carnivores, herbivores, or omnivores?
3. What does a raven eat?
4. What are the predators of ravens?
5. Does the male raven or the female raven take care of their young?
6. How do ravens fly during mating season?
7. Do ravens mate for life?

## Chapter 20
### Airplanes

*The airplane stays up because it doesn't have time to fall.*
*(Orville Wright)*

## Vocabulary

**Drag:** a force that opposes the motion of planes due to friction

**Gravity (Weight) :** an invisible force that pulls all objects towards the center of the earth

**Inventor:** someone who makes new things

**Lift:** a force that opposes gravity and keeps planes in the air

**Thrust:** a force that pushes planes through the air

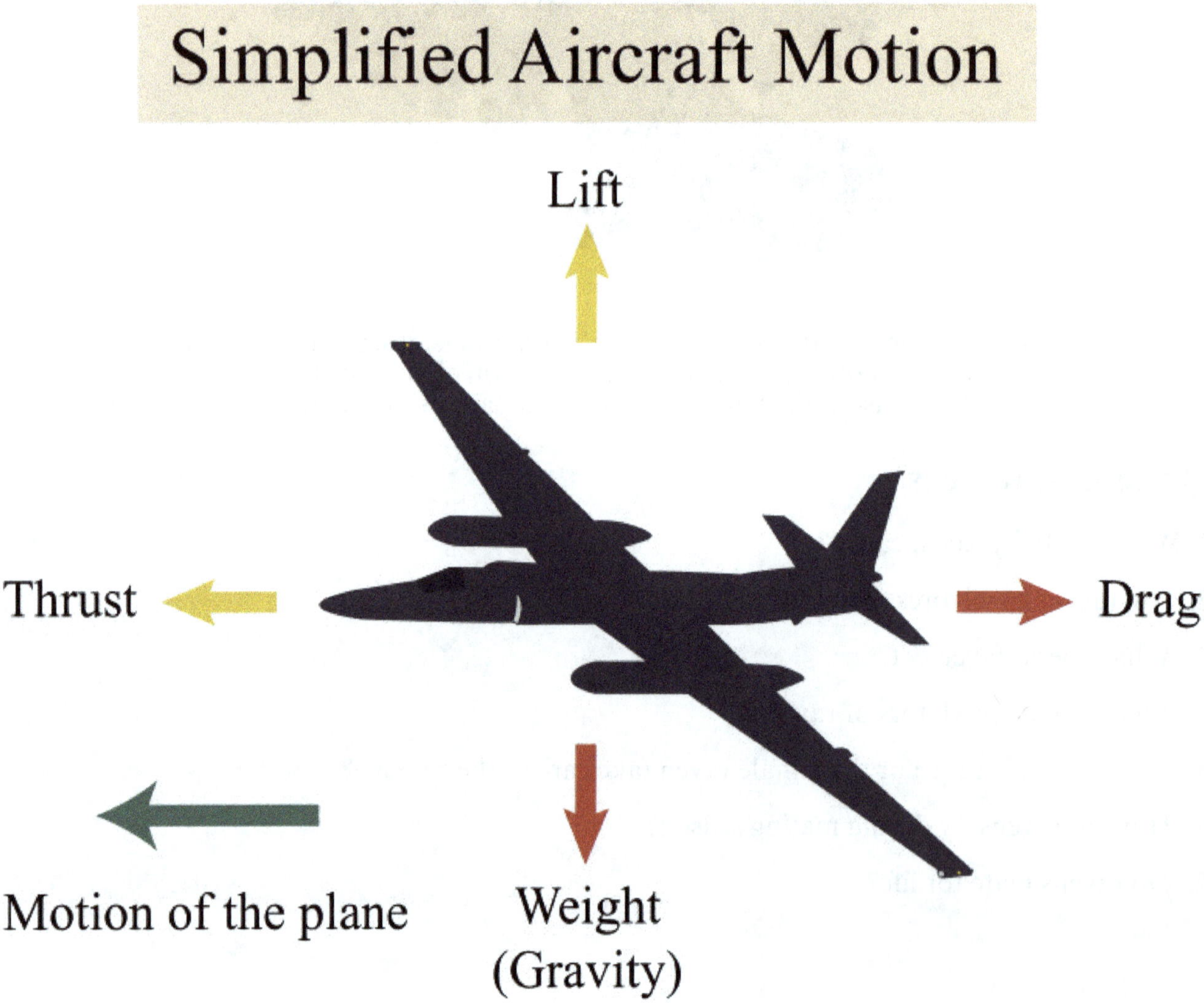

# History of the Wright Brothers

Wilbur and Orville Wright observed how different birds could roll left or right to remain stable when the winds changed. As they studied soaring birds, such as turkey vultures and hawks, the Wright brothers noticed the slight motions which enabled these birds to stay aloft without flapping their wings.

On the morning of December 17, 1903, Wilbur and Orville Wright took turns piloting their airplane, *The Wright Flier*, at Kitty Hawk, North Carolina. Orville piloted the first flight which lasted 12 seconds and reached 120 feet. The longest flight that day lasted 59 seconds and reached 852 feet.

**Boeing 787**

*Today, a Boeing 787 can fly 10,000 miles on a single tank of gas. Surprisingly, even with today's technological advances, airplanes still use the same principles of aerodynamics used by the Wright brothers.*

# Lift
## Activity

**Purpose**
To discover how airplanes fly as we think about "lift"

**Materials**
Ping pong balls
Hair dryer

**Procedure**
Turn the hair dryer on at the coolest temperature.
Put the ping pong ball into the air flow.

**Result**
The ping pong ball lifts into the airflow, demonstrating lift in an airplane.

**Why?**
Lift is a force which opposes gravity (weight) and keeps planes in the air.

# Drag
# Activity

**Purpose**
To discover how airplanes fly as we think about "drag"

**Materials**
Hand
Air

**Procedure**
Hold your hand with your thumb on top and your pinkie facing the floor.
Swing your hand back and forth. Do you feel the drag?
Turn your palm parallel to the ground; swing it back and forth.
Do you feel your hand slice through the air with less drag?

**Result**
When your hand swings palm-side down, there is less drag.

**Why?**
Wind resistance causes friction which causes drag. The same is true for airplanes.
An airplane must be designed to lessen the friction, so it can fly.

## Airplanes

# Questions: Airplanes

1. What are the four forces to consider in flight?
2. What is an inventor?
3. Who were the two brothers who invented the first airplane?
4. What force opposes gravity and keeps a plane in the air?
5. What force moves a plane through the air?
6. What force slows down movement with friction from the air (wind resistance)?
7. What force pulls objects towards the center of the earth?

# Airplanes
## Lift, Drag, Gravity (Weight), and Thrust
### Activity

**Purpose**
To discover how airplanes fly

**Materials**
Paper
Pencils, markers, or crayons
Prizes for contest winners
- Best Distance
- Best Decorated

**Procedure**
Make paper airplanes. (See supplementary material)
The students stand in a row and put their airplanes to flight.
The plane that flies the farthest gets a prize.
The plane decorated the best gets a prize.

**Result**
A paper airplane demonstrates the forces of gravity, thrust, lift, and drag.

**Why?**
Lift: This is the force that opposes gravity and holds it in the air.
Drag: The airplane with the least amount of drag will fly the farthest.
Gravity: What goes up must come down.
Thrust: The airplane moves forward with a hand thrust.

# Unit 6
## Animals and People

# Day 6
## Animals and People

*Then God said, "Let the earth produce living creatures according to their kind: livestock and crawling things and animals of the earth according to their kind"; and it was so. God made the animals of the earth according to their kind, and the livestock according to their kind, and everything that crawls on the ground according to its kind; and God saw that it was good.*

*God created man in His own image, in the image of God*
*He created him; male and female He created them.*

*God saw all that He had made, and behold, it was very good.*
*And there was evening and there was morning, the sixth day.*
*(Genesis 1:24-25, 27, 31)*

# Chapter 21
## Mammals

*Then God said, "Let the earth produce living creatures according to their kind: livestock and crawling things and animals of the earth according to their kind"; and it was so.*
*(Genesis 1:24)*

## Vocabulary

**Carnivore:** a meat-eating animal

**Ewe:** a female sheep

**Herbivore:** a plant-eating animal

**Lamb:** a baby sheep

**Mammal:** a warm-blooded vertebrate that has hair and nurses; most give birth to live young

**Omnivore:** a plant and meat-eating animal

**Predator:** an animal hunting another animal

**Prey:** an animal hunted by another animal

**Pup:** a baby wolf

**Ram:** a male sheep

**Vertebrate:** an animal with a backbone

**Warm-blooded:** an animal that maintains a constant body temperature

## Exception to the Rule

A duck-billed platypus and a spiny anteater are mammals that lay eggs.

**Duck-billed Platypus**

**Spiny Anteater**

## Questions: Mammels

1. Are mammals invertebrates or vertebrates?
2. Are mammals warm-blooded or cold-blooded?
3. Do most mammals give birth to live young or lay eggs?
4. What is a meat-eating mammal called?
5. What is a plant-eating mammal called?
6. What is a meat and plant-eating animal called?
7. What is the name for a hunted animal?
8. What is the name of the hunter of an animal?
9. Do mammals have scales or hair?
10. How do mammals feed their young?

# Wolves

*The wolf will live with the lamb…*
*(Isaiah 11:6)*

Wolves are mammals at the top of the food chain. Their only predator is humans. They have sharp teeth and claws and tend to hunt in packs. Wolves are carnivores that feast on beaver, caribou, deer, elk, mice, moose, rabbit, and squirrels. In the wild, wolves can live from 5-12 years. In zoos, they can live up to 20 years

Wolves live in deserts, forests, grasslands, mountains, and the tundra. They tend to mate for life, unless one mate dies. Their babies are called pups.

## Questions: Wolves

1. Wolves are in what animal classification?

2. What are baby wolves called?

3. Are wolves vertebrates or invertebrates?

4. Are wolves warm-blooded or cold-blooded?

5. Where do wolves live?

6. Are wolves carnivores, herbivores, or omnivores?

7. What do wolves eat?

8. What are the predators of wolves?

9. What protection did God give wolves?

10. What is the lifespan of a wolf?

11. Do wolves mate for life?

# Sheep

*When all the flocks were gathered there, they would roll the stone from the mouth of the well and water the sheep. Then they would put the stone back in its place on the mouth of the well.*
*(Genesis 29:3)*

Sheep live in grasslands. As herbivores, they eat grass and vegetation. Sheep are mammals with many predators, including bears, bobcats, coyotes, dogs, eagles, lions, lynxes, wild pigs, and wolves. Llamas or mules are often mixed in with sheep herds to ward off predators. King David is an example of a shepherd, who protected sheep from lions and bears as a young boy.

Male sheep fight for the right to mate with female sheep; the strongest is usually the only one allowed to mate. Baby sheep are called lambs. A female sheep is called a ewe. A male sheep is called a ram.

## Questions: Sheep

1. Sheep are in what animal classification?
2. What are baby sheep called?
3. Are sheep vertebrates or invertebrates?
4. Are sheep warm-blooded or cold-blooded?
5. Where do sheep live?
6. Are sheep carnivores, herbivores, or omnivores?
7. What do sheep eat?
8. What are the predators of sheep?
9. In what ways do humans help protect sheep from predators?

# Chapter 22
## Arthropods

*Of them you may eat: the locust of any kind, the bald locust of any kind, the cricket of any kind, and the grasshopper of any kind.*
*(Leviticus 11:22)*

## Vocabulary

**Arachnid:** an animal with eight legs and two body parts (daddy longlegs, scorpions, and spiders)

**Arthropod:** a cold-blooded invertebrate with an exoskeleton, segmented body parts, and jointed appendages; most lay eggs (arachnids, crustaceans, insects, and myriapods )

**Crustacean:** a sea animal with hard coverings (lobsters and crabs)

**Exoskeleton:** the hard outer covering that supports and protects arthropods

**Insect:** an animal with six legs, exoskeleton, three body parts, two antennae, and wings

**Metamorphosis:** developmental changes (caterpillar to a butterfly; tadpole to a frog)

**Molting:** shedding old feathers, hair, scales, or skin to make way for new growth

**Myriapod:** a wormlike animal with many legs (centipedes and millipedes)

**Spider:** an arachnid with  fangs that produce venom; often spins webs of silk to catch prey

## Spider Web

# Arthropods

Arthropods can be carnivores, herbivores, or omnivores, depending on the species.

*They live EVERYWHERE!*

## Questions: Arthropods

1. Are arthropods invertebrates or vertebrates?
2. Are arthropods warm-blooded or cold-blooded?
3. Where do arthropods live?
4. Are arthropods carnivores, herbivores, or omnivores?
5. What are some characteristics of arthropods?
6. Do MOST arthropods give birth to live young or lay eggs?
7. What is an exoskeleton?
8. What is molting?
9. What is a myriapod?
10. What is a crustacean?
11. What is an arachnid?
12. What is an insect?

# Lobsters

Lobsters are crustaceans with a hard outer shell. They are carnivores eating fish, crabs, clams, mussels, sea urchins, and other lobsters. Dog fish, skates, cod, and humans eat them. Most get eaten, but lobsters can live over 100 years. They grow by molting. If a lobster loses a claw, it will grow back.

## Questions: Lobsters

1. Are lobsters carnivores, herbivores, or omnivores?
2. What do lobsters eat?
3. What are the predators of lobsters?
4. What protection did God give lobsters?
5. What happens if a lobster loses one of its claws?
6. What is the lifespan of a lobster?
7. How do lobsters grow?
8. Are lobsters myriapods, crustaceans, arachnids, or insects?

# Moths and Butterflies

*But store up for yourselves treasures in heaven, where neither moth nor rust destroys, and where thieves do not break in or steal.*
*Matthew 6:20*

| *Insects* | Butterflies | Moths |
|---|---|---|
| **Feeding Behavior** | Proboscis sucks nectar from flowers. | No proboscis; survives off fat stored during the larvae stage. |
| **Wings** | Thin; scales on wings | Thick; fuzzy scales on wings |
| **Resting Position** | Wings together like praying hands | Wings lie open |
| **Temperature** | Warmed by the sun | Moves wings to warm |
| **Sound Detection** | Hears vibrations with wings | Hears sounds with ears |
| **Pupa Stage** | Chrysalis | Cocoon |
| **Activity** | Active in day | Active at night |
| **Mating** | Uses sight to select mates | Uses scent to select mates |
| **Antennae** | Long and slender with knobs at the ends | Feathery |
| **Colors** | Bright colors (most) | Dull colors (most) |

## Butterflies Compared to Moths

## Metamorphosis of Butterflies and Moths

Egg, larva (caterpillar), pupa (chrysalis/cocoon), and adult

## Questions: Butterflies and Moths

1. What is metamorphosis?
2. According to the chart on page 156, is a butterfly active in the day or the night?
3. According to the chart on page 156, who spins a cocoon, a moth or a butterfly?
4. Are moths and butterflies myriapods, crustaceans, arachnids, or insects?

**Monarch Butterfly**

**Lepidoptera Moth**

# Centipedes and Millipedes

Centipede

Millipede

| *Myriapods* | Centipedes | Millipedes |
|---|---|---|
| **Legs** | 15-30 legs – a leg lost will grow back | 47-197 legs – a leg lost will grow back |
| **Food** | Small bugs (silverfish, spiders, and ants) | Decaying matter (fruits, vegetables, and plants) |
| **Predators** | Birds, toads, frogs, and small mammals | Birds, toads, frogs, and small mammals |
| **Bite** | Poisonous bite like bee sting | Harmless |

## Questions: Centipedes and Millipedes

1. According to the chart, how many legs does a centipede have?
2. According to the chart, how many legs does a millipede have?
3. According to the chart, what are the predators of centipedes and millipedes?
4. According to the chart, what do centipedes and millipedes eat?
5. According to the chart, do centipedes or millipedes have a poisonous bite?
6. According to the chart, what happens if a centipede or millipede loses a leg?
7. Are centipedes and millipedes myriapods, crustaceans, arachnids, or insects?

# Tarantulas

A tarantula is classified as an arachnid. It has eight legs and two body parts (head and abdomen). Instead of weaving webs, this spider grabs its prey with its legs, injects paralyzing venom, and then bites with its fangs. Tarantulas secrete digestive enzymes to liquefy their victims' bodies so they can suck them through straw-like mouth openings. They eat grasshoppers, beetles, other small spiders, and sometimes even small lizards. Its venom has low toxicity, which means it is not life-threatening to humans.

The Pepsis wasp is the tarantula's greatest enemy. It can paralyze a tarantula with its sting. The wasp lays eggs on the spider's body. When the larvae hatch, they feed on the tarantula's body. Eventually, the tarantula dies.

## Questions: Tarantulas

1. How many legs does a tarantula have?
2. How many body parts does a tarantula have?
3. Is a tarantula's venom life-threatening to humans?
4. What is the main predator of a tarantula?
5. What do tarantulas eat?
6. Does a tarantula spin a web like other spiders?
7. Are tarantulas myriapods, crustaceans, arachnids, or insects?

## Tarantula
### Activity

Copy and color the tarantula.

# Chapter 23
## Reptiles

*For every species of beasts and birds, of reptiles and creatures of the sea, is tamed and has been tamed by the human race.*
*(James 3:7)*

## Vocabulary

**Carnivore:** a meat-eating animal

**Cold-blooded:** an animal that gains warmth from the environment to maintain body temperature

**Flippers:** the body parts on the sides of a sea turtles that help with swimming

**Herbivore:** a plant-eating animal

**Molting:** the shedding of old feathers, hair, scales, or skin for new growth

**Scales:** plates made of keratin, like hair, and found in an overlapping arrangement on the skin of a reptile

# Reptiles

Reptiles are cold-blooded vertebrates. Most are carnivores, but some are herbivores. Most have scales. About 600 species are venomous, and only about 200 (7% of all reptiles) are able to kill or significantly wound a human.

What is the difference between poison and venom?
Poison is delivered by swallowing, inhaling, or through the skin.
Venom is injected through a bite or a sting.

## Exception to the Rule

Most snakes lay eggs, but some bear live young.

**Boa**

**Anaconda**

**Rattlesnake**

## Questions: Reptiles

1. Are reptiles vertebrates or invertebrates?
2. Are reptiles warm-blooded or cold-blooded?
3. Do MOST reptiles give birth to live young or lay eggs?
4. Are MOST reptiles carnivores, herbivores, or omnivores?
5. What are the bony plates that protect the skin of MOST reptiles called?
6. What is the difference between poison and venom?

# Sea Turtles and Tortoises

| *Reptiles* | Sea Turtles | Tortoises |
|---|---|---|
| **Habitat** | Lakes, oceans, ponds, rivers, swamps, and wetlands | Deserts, grasslands, scrubs, and tropical forests |
| **Food** | Omnivore: algae, crabs, fish, jellyfish, seagrasses, seaweed, shrimp, and sponges | Clover, fungi, grasses, millipedes, mulberry leaves, and other omnivores |
| **Predators** | Sharks and humans; hatchlings die by birds, crabs, fish, and octopus, | Coyotes, foxes, ravens, and roadrunners |
| **Life-span** | Up to 100 years old | Up to 150 years and beyond |
| **Teeth** | None; have beaks | None; have beaks |
| **Eggs** | Lay eggs in sandy areas | Typically lay in sandy areas, but may lay eggs in dens, or under objects, such as logs |
| **Protection** | Hard shells | Hard shells |
| **Colors** | Olive-green, yellow, greenish-brown, reddish-brown, or black | Yellowish to dark brown with more yellow on their heads |

**Sea Turtle**

**Tortoise**

## Sea Turtle
### Activity

Complete and color the sea turtle.

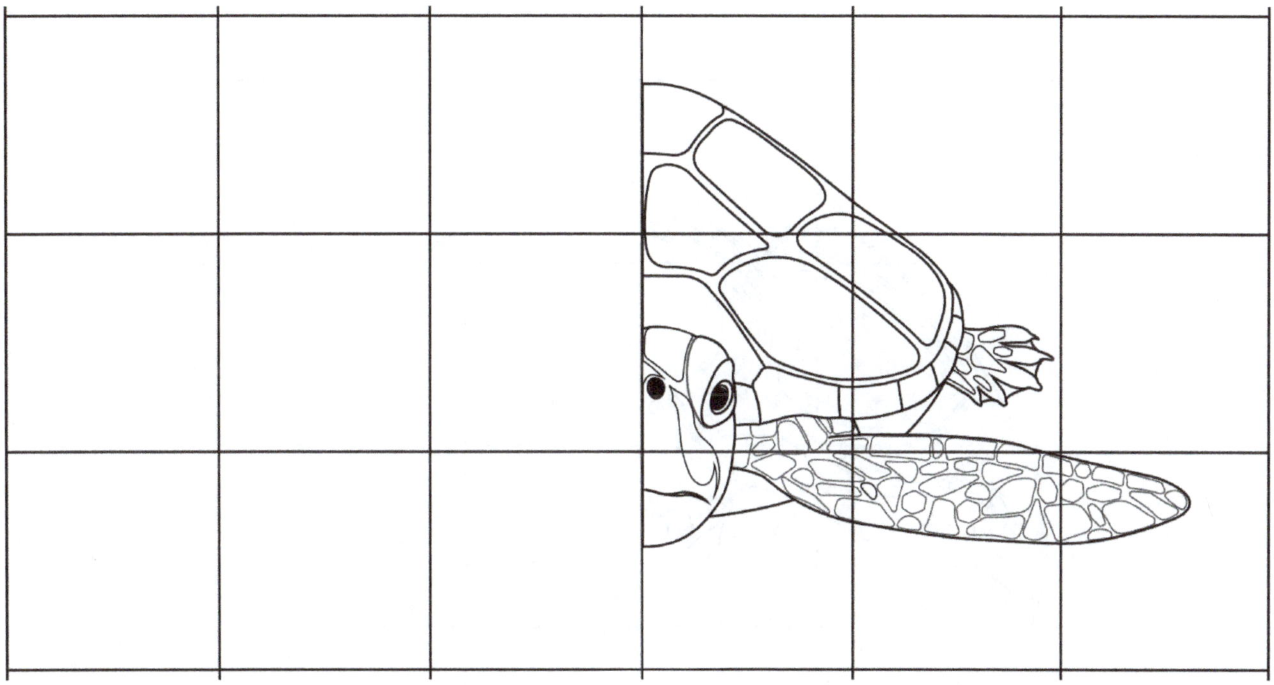

# Tortoise
## Activity

Color the tortoise.

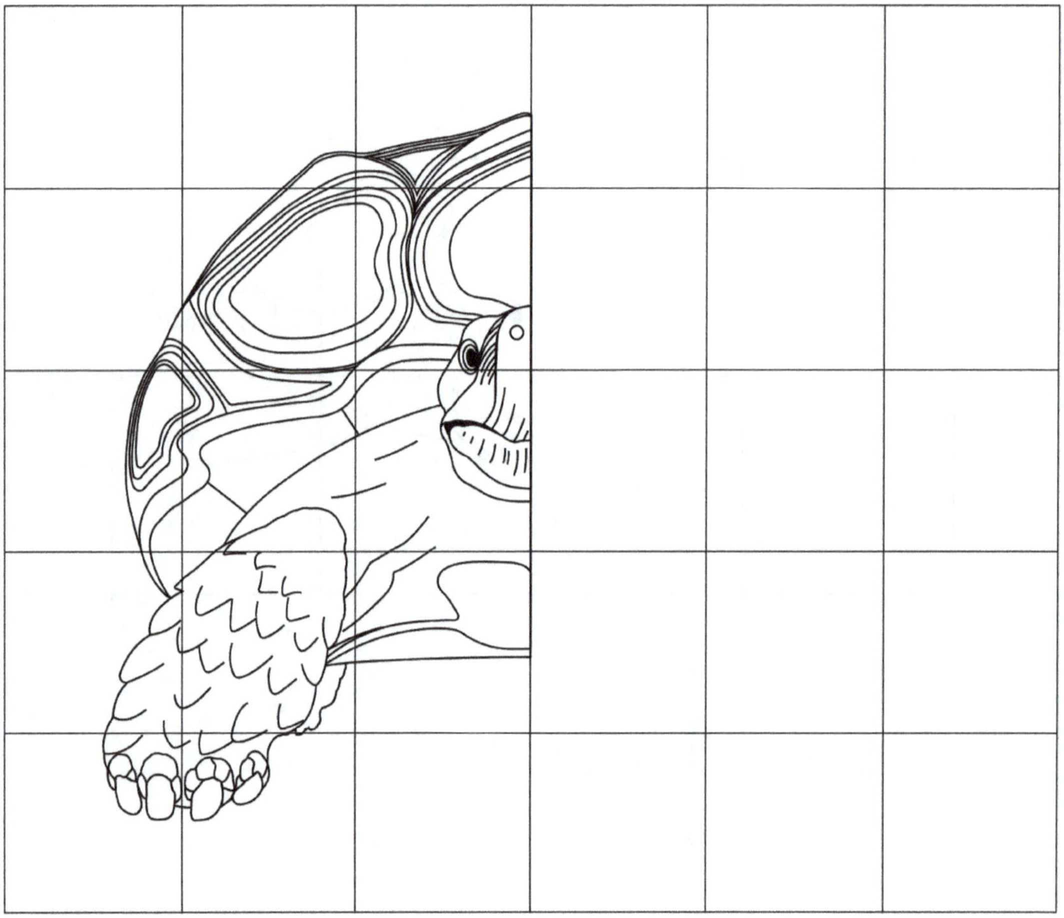

## Questions: Sea Turtles and Tortoises

*(See chart on page 163)*

1. What are the habitats of a sea turtle and a tortoises?
2. Are sea turtles and tortoises carnivores, herbivores, or omnivores?
3. What do sea turtles and tortoises eat?
4. What are the predators of sea turtles and tortoises?
5. How long can sea turtles and tortoises live?
6. Do sea turtles and tortoises have teeth?
7. Where do sea turtles and tortoises lay their eggs?
8. What protection did God give sea turtles and tortoises?

## Exception to the Rule

Most snakes have scales, but some snakes are scaleless.

### Corn Snake

Snakes are on every continent except Antarctica. Snakes use forked tongues to smell as they hunt prey. As carnivores, they eat amphibians, birds, fish, small mammals, and other snakes. Birds, boars, coyotes, foxes, mongooses, and raccoons eat them. In order to protect themselves, snakes can slither away or use defensive coiling. Their unique colors help them camouflage and remain hidden. Eighty-five percent of snakes are not poisonous.

## Questions: Snakes

1. What kind of tongue does a snake have?
2. How does a snake smell?
3. Where do snakes live?
4. Are all snakes poisonous?
5. Are snakes carnivores, herbivores, or omnivores?
6. What do snakes eat?
7. What are predators of snakes?
8. What protection did God give snakes?
9. Do MOST snakes have scales?

# Alligators and Crocodiles

| *Reptiles* | Alligator | Crocodile |
|---|---|---|
| **Habitat** | United States and China; freshwater | Every continent except Antarctica and Europe: freshwater and saltwater |
| **Food** | Carnivore: fish, deer, frogs, birds, and buffalo | Carnivore: fish, deer, frogs, birds, and buffalo |
| **Predators** | Adult: humans; Hatchlings: raccoons, birds, and crabs | Adult: humans; Hatchlings: raccoons, birds, and crabs |
| **Danger** | Mostly afraid of people | Very aggressive |
| **Size** | Smaller than crocodiles; typically, 6-12 feet long | Larger than alligators; average up to 17 feet |
| **Head** | Snout is wide and in the shape of a U; top jaw teeth can be seen. | Snout is slender and in the shape of a V; both sets of teeth can be seen. |
| **Family** | Young alligators remain in the area where they are hatched. Their mother protects them. After two to three years, they leave in search of food, or when driven out by larger alligators. | Surprisingly good parents and can be quite gentle with each other when courting, which is unusual for reptiles. Mothers take care of their babies for about three months. |
| **How long can they stay underwater?** | Nearly an hour | Nearly an hour |
| **Finding Food and Protection** | Long and powerful tails; strong jaws and swimming skills; excellent hearing, eyesight, and sense of smell | Long and powerful tails; strong jaws and swimming skills; excellent hearing, eyesight, and sense of smell |
| **Water or Land?** | Mostly in the water. Can go on land for short distances. | Mostly in the water. Can go on land for short distances. |

## Crocodile
### Activity

Complete and color the crocodile.

## Baby Crocodile

*The only place in the world to find both alligators and crocodiles is the Florida Everglades.*

### Questions: Alligators and Crocodiles

*(See chart on page 168)*

1. What skills do alligators and crocodiles have for hunting prey?
2. Do alligators and crocodiles go on land?
3. Are alligators and crocodiles carnivores, herbivores, or omnivores?
4. What do alligators and crocodiles eat?
5. What are the predators of alligators and crocodiles?
6. How long can alligators and crocodiles hold their breath?
7. Do alligators and crocodiles take care of their young?
8. Do alligators and crocodiles live in freshwater or saltwater?
9. Which is larger, an alligator or a crocodile?

11. What is the difference between the snout of an alligator and the snout of a crocodile?

# Chapter 24
## Amphibians

*Then the Lord said to Moses, "Go to Pharaoh and say to him, 'This is what the Lord says: "Let My people go, so that they may serve Me. But if you refuse to let them go, behold, I am going to strike your entire territory with frogs.*
*(Exodus 8:1-2)*

### Vocabulary

**Amphibian:** an animal who lives on land and water (frogs, toads, and salamanders)

**Carnivore:** a meat-eating animal

**Gills:** the part of the body that amphibians (and fish) use to breathe oxygen from water

**Lungs:** the organs used for breathing oxygen in the air

**Metamorphosis:** developmental changes (tadpole to a frog)

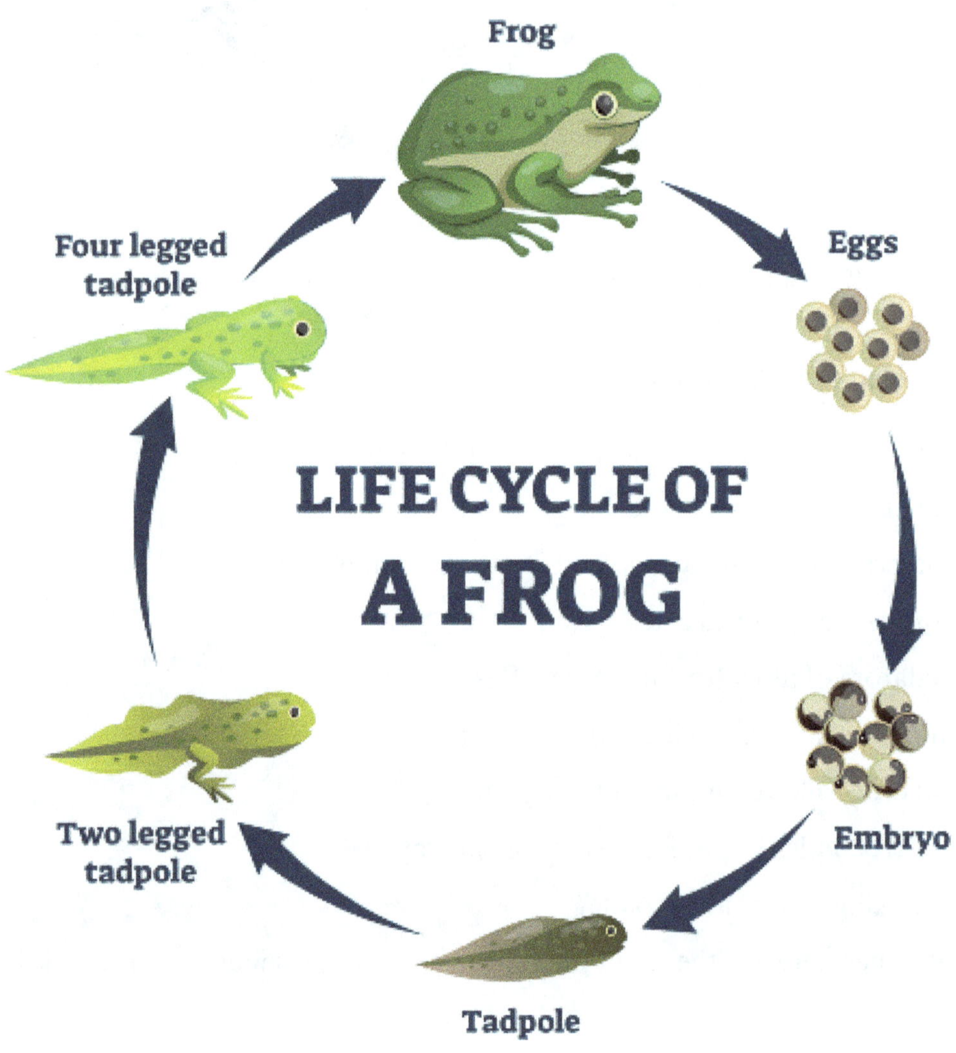

# Frogs and Toads

| *Amphibians* | Frogs | Toads |
|---|---|---|
| **Habitat** | Live mostly in water | Live mostly on land |
| **Food** | Carnivores: most eat insects, slugs, spiders, and worms; larger species eat amphibians, birds, mice, small reptiles | Carnivores: most eat insects, slugs, spiders, and worms; larger species eat amphibians, birds, mice, and small reptiles |
| **Predators** | Birds, foxes, and snakes | Birds, raccoons, and snakes |
| **Baby Name** | Tadpole | Tadpole |
| **Eggs** | Lay near or in water | Lay near or in water |
| **Lifespan** | Lives an average of 7-14 years | Lives an average of 7-14 years |
| **Legs** | Long, powerful back legs-great for jumping | Short legs |
| **Skin** | Moist and smooth | Dry and bumpy |
| **Mouth** | Beaks instead of teeth | Beaks instead of teeth |
| **Catch Prey** | Long, sticky tongue | Long, sticky tongue |
| **Metamorphosis** | Eggs, embryo, tadpole, two legs, four legs, frog | Eggs, embryo, tadpole, two legs, four legs, toad |

## Frog

## Toad

## Tadpoles in a Pond

## Questions: Frogs and Toads

*(See chart on page 172)*

1. What are baby frogs and toads called?
2. Are frogs and toads carnivores, herbivores, or omnivores?
3. What do frogs and toads eat?
4. How do frogs and toads catch prey?
5. Where do frogs and toads lay eggs?
6. What are the predators of frogs and toads?
7. What is the lifespan of frogs and toads?
8. What is the difference between the back legs of a frog or toad?
9. What is the difference between the skin of a frog or toad?
10. Where do adult frogs and toads live?
11. Do frogs and toads have teeth?

## Salamander

| *Amphibian* | **Salamander** |
|---|---|
| **Habitat** | Land and freshwater habitats in temperate and tropical forests. |
| **Food** | Carnivores: insects, flies, mosquito larvae, salamanders, slugs, spiders, and worms |
| **Predators** | Raccoons, skunks, snakes turtles |
| **Baby Name** | Eft |
| **Eggs** | Attached to stones or debris in underwater crevices |
| **Lifespan** | More than 70 years in zoos |
| **Catch Prey** | Long, sticky tongue |
| **Metamorphosis** | Eggs, larva with gill buds, larva with 2 legs, larva with 4 legs, adult |

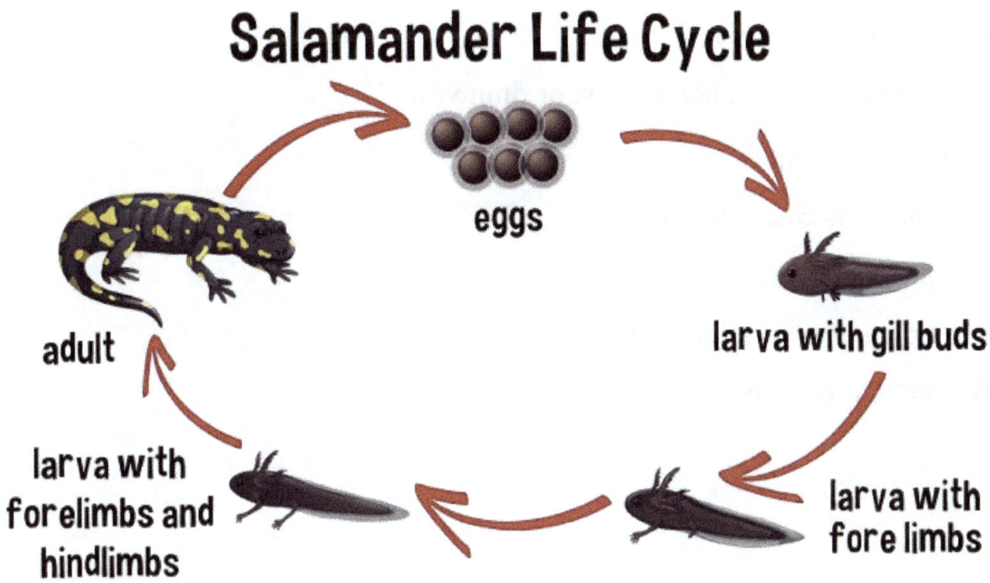

# Salamander

## Questions: Salamanders

*(See chart on page 174)*

1. What are baby salamanders called?
2. Are salamanders carnivores, herbivores, or omnivores?
3. What do salamanders eat?
4. How do salamanders catch their prey?
5. Where do salamanders lay eggs?
6. What are the predators of salamanders?
7. What is the metamorphosis of a salamander?

# AMPHIBIANS
## WORD SEARCH

| A | S | I | S | O | H | P | R | O | M | A | T | E | M |
|---|---|---|---|---|---|---|---|---|---|---|---|---|---|
| F | G | H | J | K | L | Z | X | G | C | V | B | N | P |
| Q | W | E | R | T | Y | U | I | I | O | P | C | R | V |
| B | N | M | A | S | D | L | F | H | G | J | E | K | L |
| W | E | F | R | O | G | R | T | I | E | Y | U | I | O |
| S | G | G | E | A | S | D | L | F | R | G | H | J | K |
| R | T | A | D | P | O | L | E | R | O | T | S | Y | U |
| O | H | G | N | F | S | D | S | A | V | P | G | O | I |
| T | J | K | A | M | P | H | I | B | I | A | N | L | P |
| A | S | A | M | W | E | R | T | U | N | Y | U | I | S |
| D | F | D | A | O | T | G | H | J | R | K | L | N | L |
| E | O | I | L | U | Y | T | R | E | A | W | X | C | L |
| R | P | M | A | N | B | V | C | X | C | A | S | D | I |
| P | R | T | S | Y | U | I | O | P | L | K | J | H | G |

| SALAMANDER | FROG | TOAD |
|---|---|---|
| AMPHIBIAN | TADPOLE | EGGS |
| LUNGS | METAMORPHOSIS | PREDATOR |
| CARNIVORE | GILLS | PREY |

# Chapter 25
## See

*As Jesus went on from there, two men who were blind followed Him, crying out, "Have mercy on us, Son of David!" and after He entered the house, the men who were blind came up to Him, and Jesus said to them, "Do you believe that I am able to do this?" They said to Him, "Yes, Lord." Then He touched their eyes, saying, "It shall be done for you according to your faith."*
*(Matthew 9:27-29)*

*But if your eye is bad, your whole body will be full of darkness. So, if the light that is in you is darkness, how great is the darkness! No one can serve two masters; for either he will hate the one and love the other, or he will be devoted to one and despise the other. You cannot serve God and wealth.*
*(Matthew 6:23-24)*

## Vocabulary

**Blind:** unable to see

**Braille:** an alphabet for the blind

**Braille, Louis:** invented an alphabet for the blind

**Color blindness:** unable to see certain colors

### Seeing Eye Dog

# Braille
## Activity

**Purpose**
To learn how blind people read and write

**Materials**
3x5 cards
White glue
Braille alphabet (page 179)
Markers or pencils

**Method**
Print first names in large letters on 3x5 cards.
Underneath each letter, write the corresponding letter of the Braille alphabet.
Drop white glue over each dot and let the glue dry.
Read the names by feeling the Braille dots.

**Advanced game**
Take turns writing words or messages in Braille on a whiteboard.
The next turn goes to the student who guesses the words or messages correctly.

**Result**
Students will experience a little bit of what it is like to be blind.

**Why?**
Braille connects blind people to the outside world by allowing them to read.

# Louis Braille

In the early 1800's, soldiers read secret messages with their fingers. They used raised dots to prevent alerting the enemy with sound. Motivated by this idea, Louis Braille, who himself was blind, invented a system of reading and writing for those who could not see.

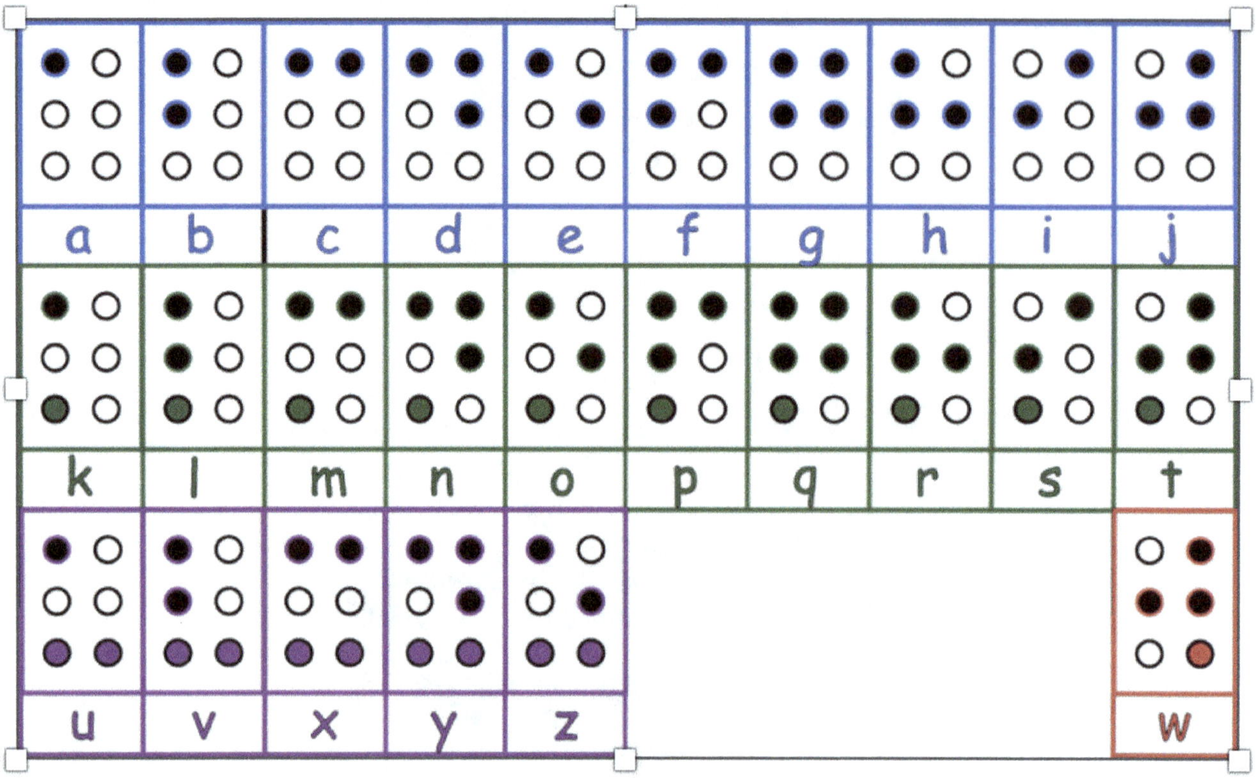

# Color Blind Test

If you are color blind, you see colors differently than most people. It usually means you have a hard time telling the difference between certain colors, not all colors.

Can you see all the numbers in the circles?

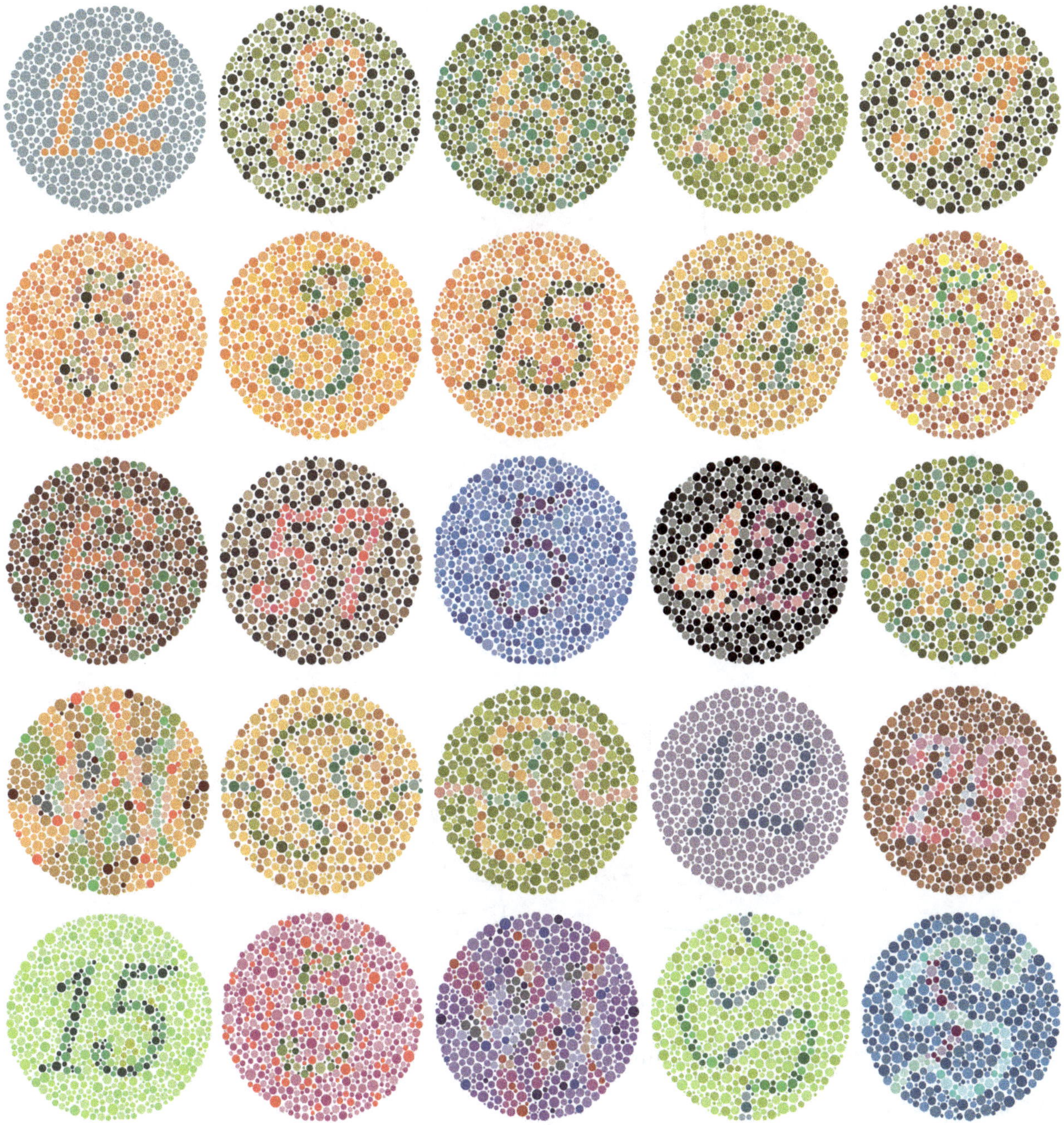

*I once was blind but now I see…*
John Newton
(Former slave trader before becoming a Christian)

## Questions: Blind

1. What alphabet was created for the blind?
2. Who created the alphabet for the blind?
3. What does it mean to be color blind?

## Activity

Create a secret message or word using the Braille alphabet.

# Chapter 26
## Hear

*And they brought to Him one who was deaf and had difficulty speaking, and they begged Him to lay His hand on him. And Jesus took him aside from the crowd, by himself, and put His fingers in his ears, and after spitting, He touched his tongue with the saliva; and looking up to heaven with a deep sigh, He said to him, "Ephphatha!" that is, "Be opened!" And his ears were opened, and the impediment of his tongue was removed, and he began speaking plainly.*
*(Mark 7:32-34)*

## Vocabulary

**Deaf:** unable to hear

**Sign language:** communication with hands used by those who are deaf

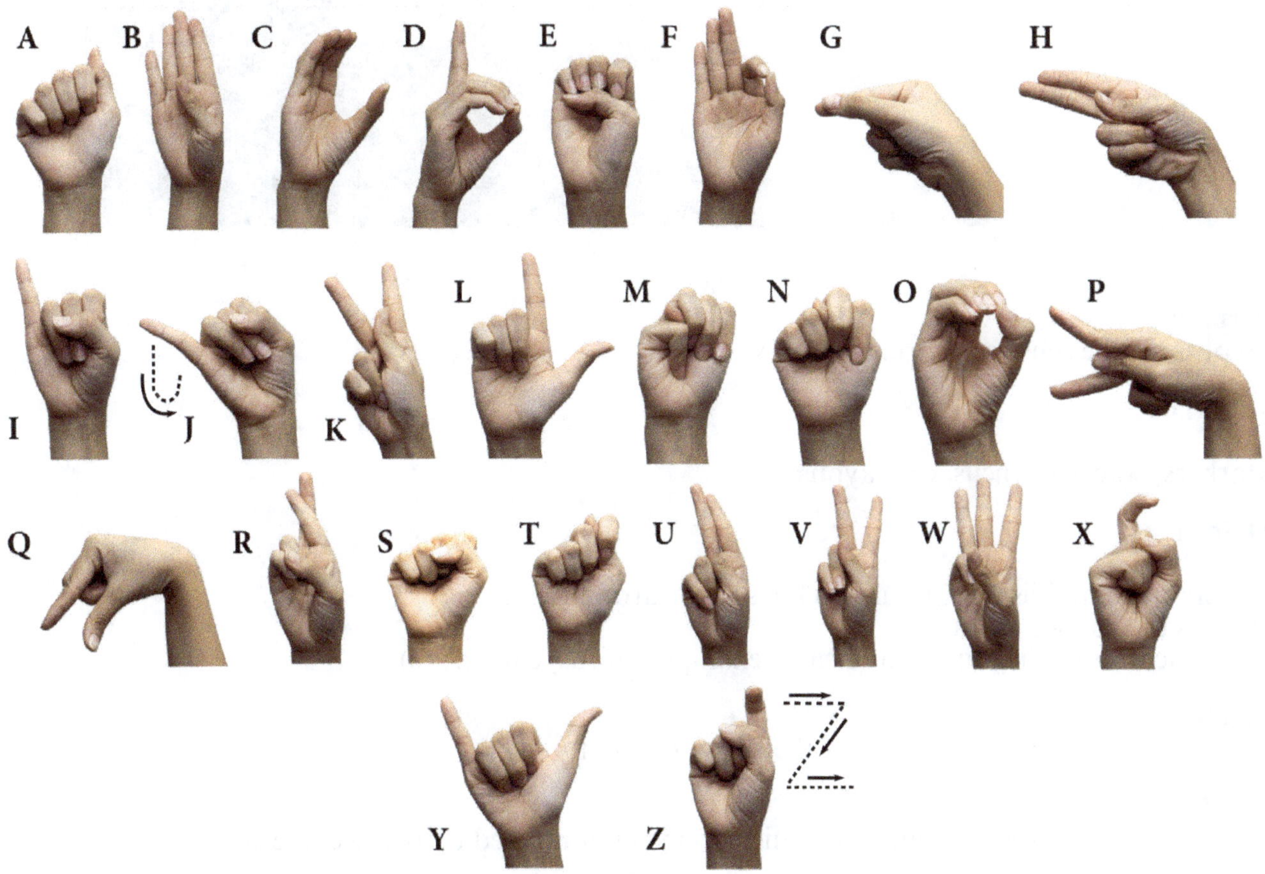

# Listening
## Activity

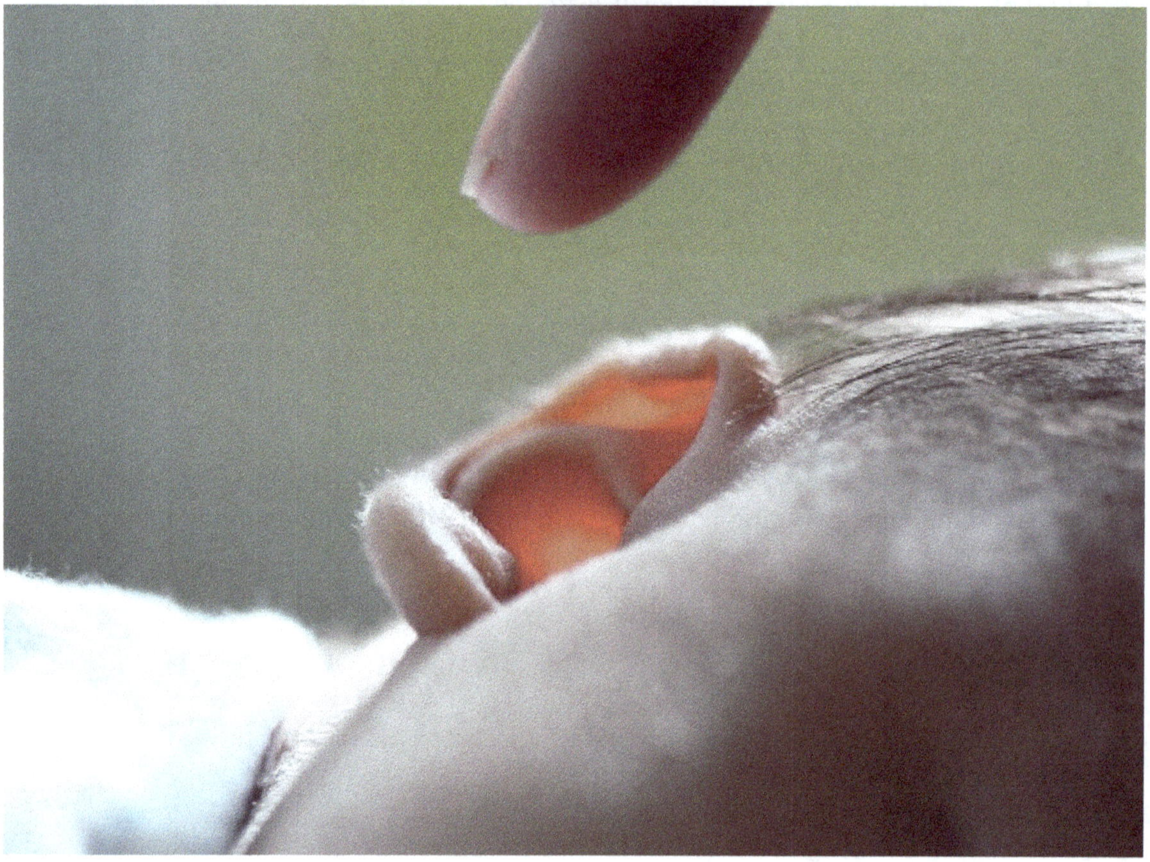

**Purpose**
To observe the sense of hearing

**Materials**
Paper
Markers, colored pencils, or crayons

**Method**
Go outside.
Be very quiet and listen carefully to the sounds around you.
Discuss the sounds in class.
Draw pictures of the things you heard (examples: birds, cars, dogs).

**Result**
The children will be more aware of their environment.

**Why?**
When someone takes the time to be quiet, sounds not noticed before become clear.

# Sign Language
## Activity

**Purpose**
To learn sign language

**Materials**
**Songs:** (YouTube)
*Jesus Loves Me*
*The Wise Man Built His House*
*You Are My All in All*

**Procedure**
Learn simple songs and their signs.

**Result**
Students will be able to communicate with the Deaf and others lacking communication skills.

**Why?**
Sign language can be used with the Deaf, babies, and those who have difficulty communicating.

# Ten Commandments
## Activity

**Purpose**
The pictures on page 186 were created for the deaf in order to teach them the Ten Commandments. Hearing children (and adults) also learn quickly by matching pictures with commandments.

**Materials**
Pictures of the Ten Commandments (see page 186)

**Procedure**
Show the pictures of the Ten Commandments.
Keep reviewing until the children are familiar with all ten commandments.
Memorize the poem.

**Result**
Most of the students will memorize the commandments surprisingly fast.
Memorization has become a lost art. Pictures make memorization easier.

## Ten Commandments

*(Exodus 20:1-17)*

Above all else love God alone.
Bow down to neither wood nor stone.
God's name refuse to take in vain.
The Sabbath rest with care maintain.
Respect your parents all your days.
Hold sacred human life always.
Be loyal to your chosen mate.
Steal nothing neither small nor great.
Report with truth your neighbor's deed.
And rid your mind of selfish greed.

# Chapter 27
## Taste and Smell

*Taste and see that the Lord is good;*
*How blessed is the man who takes refuge in Him!*
*(Psalm 34:8)*

*You shall make from these a holy anointing oil, a fragrant mixture of ointments, the work of a perfumer; it shall be a holy anointing oil.*
*(Exodus 30:25)*

## Vocabulary

**Bitter:** unsweet and sharp (coffee, unsweetened chocolate, and black tea)

**Essential oils:** natural oils from plants (lavender, peppermint, and clove)

**Salty:** saline (potato chips, crackers, and French fries)

**Savory:** pleasant, full flavor; not sweet (sauces, beef jerky, and meat)

**Sour:** acidic, not sweet (lemons, grapefruits, and limes)

**Sweet:** sugary (fruits, honey, and maple syrup)

# Taste
## Activity

**Purpose**
To discern the difference between tastes

**Materials**
IMPORTANT: Check for allergies!

Water to drink
Pencil
Paper plates
Lemons
Grapefruits
Bananas
Cookies
Coffee
Unsweetened chocolate
Potato chips
Salt water
Beef jerky
Pizza
Taste chart (below)

**Procedure**
Taste foods.
Drink water in between foods to reset taste buds.
Categorize each food under the appropriate taste on the chart below.

**Result**
The differences between tastes will be discovered.

**Why?**
One of the symptoms of Covid-19 is the loss of smell and taste.
If you cannot smell, you cannot taste. If you cannot taste, you cannot smell.

| Sour | Sweet | Bitter | Salty | Savory |
|---|---|---|---|---|
| | | | | |

# Smell
## Activity

**Purpose**
To experience the sense of smell

**Materials**
A variety of essential oils (lavender, peppermint, clove, etc.)
Paper
Pencils

**Procedure**
Fold a piece of paper in half and then in half again creating four squares.

**First square:**
Write the words "Essential Oils".

**Second square:**
Write the word "Good".

**Third square:**
Write the word "Bad".

**Fourth square:**
Write the word "Neutral".

Give a brief description of each oil.
Smell each oil.
After each smell, write the name of that essential oil on a whiteboard.
Students will copy the name of each oil under "Good," "Bad", or "Neutral".

**Uses:** (There are hundreds of essential oils – this is only a small sample.)
Lavender: soothes burns and helps with sleep
Peppermint: calms upset stomachs and relieves headaches
Clove: numbs gum pain and kills germs

**Result**
The wonder of various essential oil smells will be experienced.

**Why?**
Essential oils have been used for thousands of years for medicinal and health purposes. Better health can be achieved by using them.

# Taste and Smell Together
## Activity

**Purpose**
To observe how taste and smell work together

**Materials**
IMPORTANT: Check for allergies!

Apples: 3 chunks per student
Tomatoes: 3 chunks per student
Bananas: 3 chunks per student
Knife
Cutting board
3 Plates
Towel
Peppermint Essential Oil
Pencils
Plain Crackers
Water bottle
Cups
Toothpicks
Taste chart (page 192)

**Procedure**

**For Teacher:**
Chop and place equal amounts of fruit onto three separate plates.
Stick a toothpick into each piece.
Cover with a towel to keep the fruit hidden.

Plate 1: Put ten drops of peppermint oil onto each piece of fruit.
Plate 2: Do nothing to the fruit.
Plate 3: Do nothing to the fruit.

Blindfold and test each student separately.

**For Student:**
**Test 1:** Fruit with Peppermint Oil
While blindfolded, taste each fruit separately; drink water and eat crackers between tastes.
You have three seconds to identify the fruit.
If your identification was correct, a check-mark will be placed by that fruit on your chart.
If your identification was incorrect, a 0 will be placed by that fruit on your chart.

**Test 2:** Fruit Plain with Nose Holding
Repeat the same experiment without the peppermint oil.
While blindfolded, pinch your nose while tasting the fruit.
Observations will be recorded similar to test 1.

**Control Test 3:** Fruit without Nose Pinching
Repeat the same experiment using the third plate of fruit.
While blindfolded, simply eat the fruit and guess.
Observations will be recorded similar to test 1.

## Taste Chart

| Fruit | Peppermint Oil | Nose Plugged | Control (Plain Fruit) |
|---|---|---|---|
| Apple | | | |
| Tomato | | | |
| Banana | | | |

**Result**
Both peppermint oil and holding your nose will make it difficult to discern tastes.

**Why?**
Peppermint oil and plugging your nose confuse the sense of smell.
The sense of smell and the sense of taste work together to give correct tastes.

## Lavender Field

## Questions: Taste

1. What are some examples of bitter foods?
2. What are some examples of salty foods?
3. What are some examples of savory foods?
4. What are some examples of sour foods?
5. What are some examples of sweet foods?
6. What kind of fragrant oils come from plants?

# Taste
## Home Activity

Chart the foods you eat during the week into five taste categories.

| Sour | Sweet | Bitter | Salty | Savory |
|---|---|---|---|---|
| | | | | |

# Chapter 28
## Touch

*I will give thanks to You, because I am awesomely and wonderfully made;*
*Wonderful are Your works, and my soul knows it very well.*
*(Psalm 139:14)*

*Moved with compassion, Jesus reached out with His hand and touched him,*
*and said to him, "I am willing; be cleansed." And immediately*
*the leprosy left him, and he was cleansed.*
*(Mark 1:41-42)*

## Vocabulary

**Arch:** fingerprints where lines begin on one side, go up, and exit on the opposite side

**Fingerprints:** impressions of fingertips

**Leprosy:** a disease resulting in the loss of the sense of touch

**Loop:** fingerprints where some lines start on one side, curve upward, and go back around to the side where it began

**Whorl:** fingerprints with a circular pattern

# Fingerprints
## Activity

Loop Whorl Arch

**Purpose**
To identify different patterns of fingerprints

**Materials**
3x5 cards
Ink pads

**Procedure**
Press the top of your fingers on an ink pad; immediately press the ink on your card.
Determine if your fingerprints are arches, loops, or whorls.
Write your name and your fingerprint type (arch, loop, or whorl) on your card.

**Result**
All fingerprints will fall into one of three categories: arch, loop, or whorl.

**Why?**
Amazingly there are only three categories of fingerprints, yet EVERY fingerprint is different, including those of identical twins.

**Purpose**
To enjoy a fun activity using thumbprints

**Materials**
Different colored ink pads
Fingerprint Art: (above)
White paper

**Procedure**
Complete the animals in the picture.
Create new fingerprints on a blank sheet of paper using ink pads.
Draw original animal pictures using your own finger prints.

**Result**
The students will improve their artistic talents and have fun.

**Why?**
Just for fun!

# Fingerprints

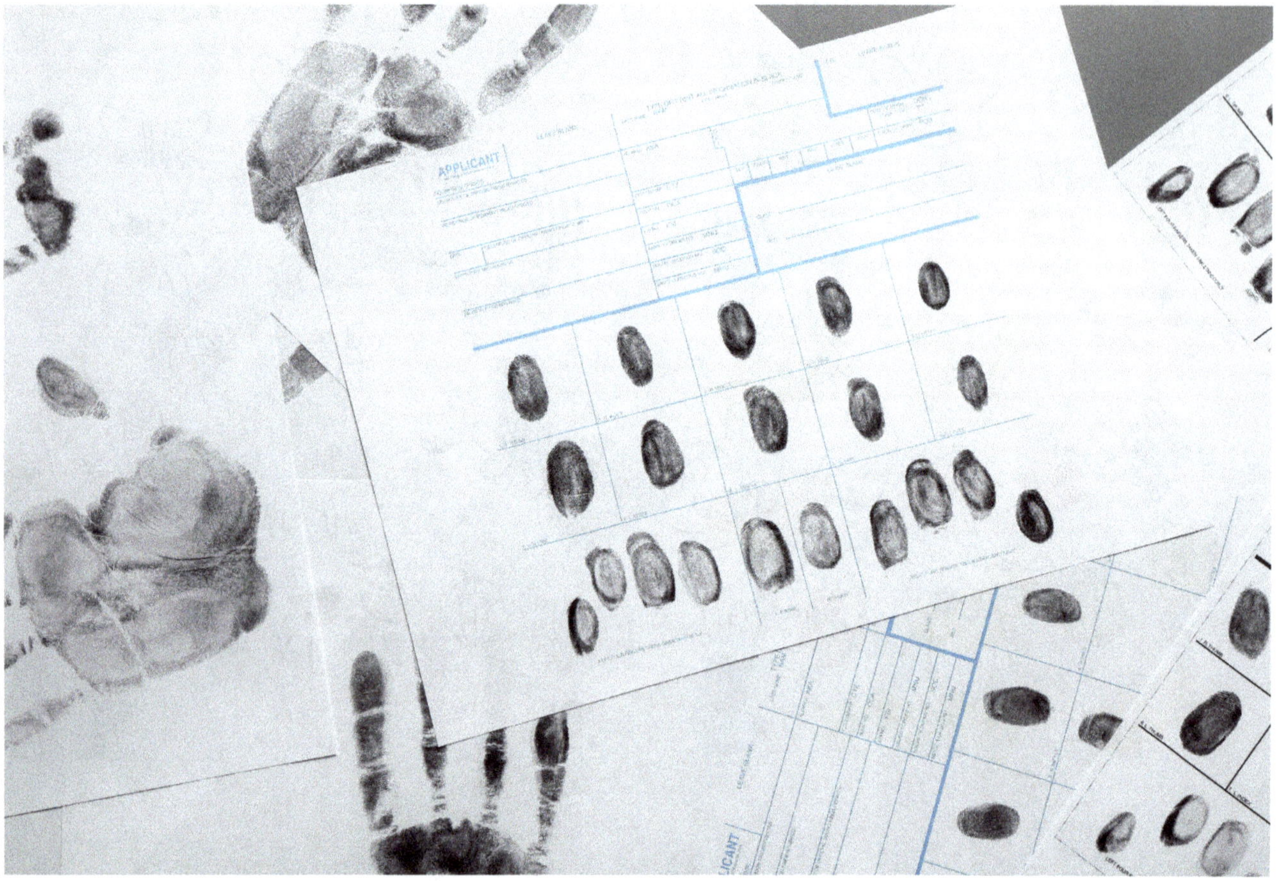


## Questions: Fingerprints

Circle the correct answer.

1. Pattern of fingerprints where some lines start on one side, curve upward, and go around to the side where it began:

**Loops, Whorls, or Arches**

2. Pattern of fingerprints where lines begin on one side, go up, and exit on the opposite side:

**Loops, Whorls, or Arches**

3. A circular pattern of fingerprints:

**Loops, Whorls, or Arches**

4. Are all fingerprints the same? Yes or No

5. Do twins have the same fingerprints? Yes or No

# Unit 7
## God Rested

*In peace I will both lie down and sleep,*
*For You alone, Lord, have me dwell in safety.*
*Psalm 4:8*

## Day 7
### God Rested

*And so, the heavens and the earth were completed, and all their heavenly lights. By the seventh day God completed His work which He had done, and He rested on the seventh day from all His work which He had done. Then God blessed the seventh day and sanctified it, because on it He rested from all His work which God had created and made.*
*(Genesis 2:1-3)*

# Chapter 29
## Rest

*Jesus said to them, "The Sabbath was made for man,*
*and not man for the Sabbath."*
*(Mark 2:27)*

## Vocabulary

**Germs:** disease-causing microscopic organisms

**Immune System:** a system of the body that fights disease

**Sabbath:** a day to worship and to rest from work; kept by Jewish people from Friday evening to Saturday evening and by many Christians all day Sunday

**Sleep:** the rest cycle

## Benefits of Sleep

**Sharper brain:** Sleep plays an important role in learning and memory.

**Mood boost:** Sleep helps prevent depression.

**Healthier heart:** Sleep gives the heart and blood vessels rest.

**Athletic performance:** Sleep improves body strength.

**Blood sugar regulation:** Sleep helps prevent Type 2 Diabetes.

**Germ-fighting:** Good sleep improves the immune system.

**Weight control:** Poor sleep causes weight gain and obesity.

## How Well Do You Sleep?

How many of the "yes" tips do you do on a regular basis?
How many of the "no" tips do you do on a regular basis?

# Sleep Noise
## Activity

**Purpose**
To learn to rate sleep noises in order of preference
To teach critical thinking

**Materials**
Colored pencils, markers, or crayons
Sleep Chart (see below)
Recordings of different sleep noises (see supplementary material)

**Procedure**
Color the circles the written colors.
Listen to the varying colors of sleep noises.
Rate the sounds: (1) most favorite sound (6) least favorite sound
Fill in (2), (3), (4), and (5), according to preference.

## Sleep Chart Noises

**_____ 0 White**
**_____ 0 Pink**
**_____ 0 Yellow**
**_____ 0 Brown**
**_____ 0 Blue**
**_____ 0 Red**

**1 2 3 4 5 6**

**Result**
Rating skills and critical thinking will be encouraged.

**Why?**
Babies, children, and adults use sound machines to sooth themselves to sleep.

## Discussion Starters

• When I have a hard time falling asleep, I can….
• When I wake up during the night, I usually…
• When I get a good night's sleep, I wake up feeling…
• When I don’t get a good night's sleep, I wake up feeling…

## Recommended Sleep

**Newborn:** 14-17 hours
**Infant:** 12-15 hours
**Toddlers:** 11-14 hours
**Preschool:** 10-13 hours
**School Age:** 9-11 hours
**Teens:** 8-10 hours
**Older Adults:** 7-8 hours

## Questions: Sleep

1. According to the chart above, how many hours should you sleep?
2. According to the chart above, how many hours should an infant sleep?
3. According to the chart above, how many hours should a teen sleep?
4. What microscopic organisms cause disease?
5. Which body system is used to fight disease?

# Recommended Sleep
## Home Activity

**Newborn:** 14-17 hours
**Infant:** 12-15 hours
**Toddlers:** 11-14 hours
**Preschool:** 10-13 hours
**School Age:** 9-11 hours
**Teens:** 8-10 hours
**Older Adults:** 7-8 hours

Record how many hours you sleep each night.

_____ **Sunday night**

_____ **Monday night**

_____ **Tuesday night**

_____ **Wednesday night**

_____ **Thursday night**

_____ **Friday night**

_____ **Saturday night**

According to the above chart, how many hours you are supposed to sleep?

_________________ hours

# Chapter 30
## Grand Finale Review

*In the beginning God created the heavens and the earth.*
*(Genesis 1:1)*

# Review Party
## Food and games!

**Purpose**
To review the days of creation in Genesis 1-2
To review all eight books of Exploring God's Creation

**Materials**
Chef hat and apron
Large knife
Plates
Chocolate sandwich cookies with white filling
Helium balloons or regular balloons
Spray bottle filled with water
Bottles of water (1 per student)
Straws
Refrigerated chocolate cookie dough: 2 rolls
Celery with leaves
Cookie sheet
Whole cantaloupe
Prepared cantaloupe chunks
Goldfish crackers
Egg
Animal cookies

# Seven Days of Creation
## Grand Finale
### Activity

**Procedure**
Give plates to each student.

**1. Day and Night**
Separate chocolate white filled sandwich cookies. The dark half represents darkness; the white half represents light.

**2. Sky and Water**
Spray each student with a water bottle. Release balloons into the air.

**3. Dry Land and Plants**
Slice cookie dough into round circles to represent dirt. Stick leafy celery tops into the "dirt" to represent plants.

**4. Sun, Moon, Stars, and Planets**
Cut a cantaloupe in two, making one half bigger than the other. The bigger half represents the sun; the smaller half represents the moon; the seeds represent the stars. Pass out cantaloups chunks to each student.

**5. Fish and Birds**
Pass goldfish crackers to eat and use an egg to represent birds.

**6. Man, Woman, and Animals**
Mold the rest of the cookie dough into a man.
Remove some of this dough off the man's side and mold it into a woman.
Serve animal cookies to represent the animals.

**7. God Rested**
Place a straw in each water bottle.
Rest, drink water, and eat food.

## Review
### Activity

**Materials**
Whiteboard and markers
Pencils
Paper for brainstorming answers
Questions from the chapters of *Exploring God's Creation*

**Set-up**
Divide the students and name the teams by numbers.

Put the team numbers on a whiteboard.

Place one chair per team in a horizontal row.

A student from each team will sit in a chair with all other team members lined behind.

**Procedure**
Ask the sitting students from multiple teams a question from one of the chapters.
The first person to stand answers the question: 3 points for the correct answer
If the answer is incorrect, the next person will try: 2 points for the correct answer
If the question is still not answered correctly, all teams will brainstorm in separate areas.
A student from each team will give an answer to the teacher: 1 point for each correct answer
The team with the most points wins.

*For smaller classes this game may be played one-on-one without teams.

**Result**
Book material is reviewed.

**Why?**
Progress and retention can be evaluated.

IN THE
BEGINNING
GOD
CREATED
THE HEAVENS AND
THE EARTH

GEN 1:1

# Exploring God's Creation

## Teacher Helps

Exploring With Grandma Publishing, Virginia



exploringwithgrandma@gmail.com
exploringwithgrandma.com

# Supplies

## Introduction

### Creation, Sin, and Rescue
Questions: Creation and Sin (page 7)
Questions: Sin (page 8)
Questions: Rescue (page 9)
Word Search (page 10)
Make Your Own Word Search (page 11)

### Chain Reactions
Dominoes

### A Firm Foundations
Wooden blocks
10-Inch string
½-Inch metal washer
Tacks
Colored pencils, markers, or crayons
Coloring Sheet: Two Paths (page 19)
Art: Cornerstone (page 20)
Questions: A Firm Foundation (page 21)

## Unit 1: Day and Night

### Reflective and Refractive Light
3x5 Index cards
Mirror
Pencil
Glass jar
Water
Questions: Refractive Light (page 26)
Activity: Reflective and refractive light (page 27)

### Colors
Questions: Flood (page 28)
Questions: Rainbow (page 29)
Coloring page: Rainbow (page 31)
Activity: Colors in the Bible (page 32)
4 Clear glasses
Food coloring (blue, red, and yellow)
Water
Colored pencils, markers, or crayons
Activity: Color wheel (page 34)
Questions: Colors (page 35)

### Light and Darkness
Activity: Light bulbs (page 37)
Glue
Black sand
Yellow colored pencils, markers, or crayons
Questions: Jesus is the Light of the World (page 38)
Art: Firefly (page 40)
Questions: Fireflies (page 41)
Questions: Jellyfish (page 43)
Activity: Light and Darkness (page 44)

### Caves, Brown Bears, and Bats
Questions: Stalactites and Stalagmites (page 46)
Questions: Brown Bear (page 48)
Art: Brown Bear (page 49)
Questions: Bat (page 51)
Art: Bat (page 52)

## Unit 2: Sky and Water

### Clouds

Coloring page: Clouds (page 56)
Questions: Clouds (page 57)

### Water Cycle

Coloring page: Water cycle (page 59)
4 Clear glasses (not patterned - 3 experiments)
Water
Permanent marker
Electric kettle
⅔ Cup hot water
4 Ice cubes
Shaving cream
Blue food coloring
Questions: Water Cycle (page 63)

### The Invisible

Guitar
Lyrics and chords
½ Cup lemon juice
Cups
3x5 Cards
Small paintbrushes
Iron
Blow dryer
Glo Germ Gel (artificial germs)
UV Flashlight
Microscope
Hair
Onion skin
Prepared slides
Paper
Drawing activity: The Invisible (page 68)
Questions: The Invisible (page 69)

### Float or Sink

Questions: Basilisk Lizard (page 72)
Dishwashing soap
Pepper
Water
White bowl
Glass jar
Egg
Salt
Questions: Float or Sink (page 74)
Home activity: Float or Sink (page 75)

## Unit 3: Dry Land and Plants

### Plants

Labeling: Parts of Plants (pages 78-79)
3 Glass flower vases (no pattern)
3 White carnations
Food colors (red, yellow, and blue)
Water
Word Search (page 81)
Questions: Plants (page 82)

### The Sower and the Seed

4 Plastic cups
Spade
Permanent markers
Fast-growing grass seeds
Dirt from a path

Dirt mixed with rocks
Dirt mixed with plant thorns
Questions: Good Soil (page 86)

### Poisonous Plants
Matching: Poisonous Plants (page 91)
Questions: Mushroom (page 92)

### Habitats
Matching: Habitats (page 97)
6 Hula hoops
Paper
Colored pencils, markers, or crayons
Numerous toy animals
Questions: Habitats (page 99)

## Unit 4: Sun, Moon, Stars, and Planets

### Sun and Other Stars
Light Source (lamp or projector)
Activity: Shadows (page 106)
Activity: Shadows (page 107)
Questions: Outer Space (page 108)

### Earth's Moon
Activity: Phases of the Moon (page 113-114)
White paper plates
Chocolate sandwich cookies with white filling (4 per student)
White glue
Scissors
Questions: Moon (page 114)

### Planets
Word Search (page 117)
Questions: Solar system (page 118)
Basketball
Tennis ball
Magnets of varying kinds and sizes
Paperclips
Questions: Isaac Newton (page 122)

## Unit 5: Fish and Birds

### Fish
Questions: Fish (page 126)
Questions: Sharks (page 127)
Questions: Bluefin Tuna (page 128)
Questions: Atlantic Salmon (page 129)
Word Search (page 130)

### Boats
Comparison: Boat and Fish (pages 132-133)
Questions: Boats (page 134)

### Birds
Milk cartons
Scissors
Paint
Buttons
Stickers
Dowel
Twine
Peanut butter
Birdseed
Questions: Birds (page 137)
Questions: Eagle (page 138)
Questions: Vulture (page 139)
Questions: Raven (page 140)

### Airplanes
Ping pong balls
Hair dryer
Paper
Pencils, markers, or crayons
Prizes for contest winners
Questions: Airplanes (page 145)

## Unit 6: Animals and People

### Mammals
Questions: Mammals (page 150)
Questions: Wolves (page 151)
Questions: Sheep (page 152)

### Arthropods
Questions: Arthropods (page 154)
Questions: Lobsters (page 155)
Questions: Butterflies and Moths (page 157)
Questions: Millipedes and Centipedes (page 158)
Paper
Colored pencils, markers, or crayons
Questions: Tarantula (page 159)
Art: Tarantulas (page 160)

### Reptiles
Questions: Reptiles (page 162)
Art: Sea turtle (page 164)
Art: Tortoise (page 165)
Questions: Sea turtles and Tortoises (page 166)
Questions: Snakes (page 167)
Crocodile Art (page 169)
Questions: Alligators and Crocodiles (page 170)

### Amphibians
Questions: Frogs and Toads (page 173)
Questions: Salamanders (page 175)
Word Search (page 176)

### See
3x5 Cards
White glue
Activity: Braille Alphabet (page 179)
Markers or pencils
Activity: Color Blind Test (page 180)
Questions: Blind (page 181)

### Hear
Paper
Markers, colored pencils, or crayons
Songs: (YouTube)
*Jesus Loves Me*
*The Wise Man Built His House*
*You Are My All in All*
Handout: Ten Commandments (page 186)

### Taste and Smell
Pencil
Water to drink
Paper plates
Lemons
Grapefruits
Bananas
Cookies
Coffee
Unsweetened chocolate
Potato chips
Salt water

Beef jerky
Pizza
Activity: Taste Chart (page 188)
A variety of essential oils (lavender, peppermint, clove, etc.)
Paper
Apples: 3 chunks per student
Tomatoes: 3 chunks per student
Bananas: 3 chunks per student
Knife
Cutting board
Plastic wrap
Peppermint essential oil
Plain crackers
Water bottle
Cups
Toothpicks
Activity: Taste chart (page 191)
Questions: Taste (page 192)
Activity: Taste chart (page 193)

### Touch

3x5 Cards
Ink pads
Different colored ink pads
White paper
Art: Thumbprint (page 196)
Questions: Fingerprints (page 197)

## Unit 7: God Rested

### Rest

Colored pencils, markers, or crayons
Recordings: Different sleep noises (YouTube)
Handout: Sleep Chart Noise (page 202)
Questions: Sleep (page 203)
Activity: Recommended Sleep (page 204)

### Grand Finale

Chef hat and apron
Large knife
Plates
Chocolate sandwich cookies with white filling
Helium balloons
Spray bottle filled with water
Bottles of water
Straws
Brown refrigerated cookie dough: 2 rolls
Celery with leaves
Cookie sheet
Whole cantaloupe
Cantaloupe chunks
Goldfish crackers
Animal cookies
Whiteboard
Markers
Pencils
Paper
Questions from Exploring God's Creation (all chapters)

# Answers

## Introduction

### **Creation and Sin** (page 7)
1. Adam and Eve disobeyed God and ate the forbidden fruit.
2. Adam and Eve had to leave the beautiful garden. Life became more difficult.

### **Sin** (page 8)
1. Cain offered fruit instead of a lamb; God was displeased.
2. Cain became angry and killed his brother. He became a fugitive and a wanderer on the earth. He could not grow crops.

### **Rescue** (page 9)
1. Jesus died on a cross so our sins can be forgiven. He rose again and is preparing a place for those who love and trust Him in heaven.
2. Satan's power has been crushed. Our sins are forgiven. Jesus is coming back again for His own people to live with Him forever.

### **The Three Little Pigs** (page 21)
1. vertical
2. horizontal
3. plumb line
4. plumb
5. cornerstone
6. foundation

## Unit 1: Day and Night

### **Reflective and Refractive Light** (page 26)
1. refraction
2. reflection

### **Flood** (page 28)
1. People were wicked. God sent a flood to destroy the whole earth.
2. Only Noah, his family, and the animals on the ark were saved from drowning.

### **Rainbow** (page 29)
1. Using refraction from a lot of water, God created a beautiful rainbow.
2. The rainbow is a covenant between God and man to never destroy the whole earth again with a flood

### **Colors in the Bible** (page 32)
1. green, freshness, rebirth
2. white, purity, holiness
3. red, blood, sin
4. purple, rich, royalty
5. gold, silver, wealth

### **Colors** (page 35)
1. blue, red, and yellow
2. green, orange, and purple
3. red, orange, yellow, green, blue, indigo, and purple
4. circle
5. never again to destroy the whole earth with a flood
6. no
7. refractive light

### **Jesus is the Light of the World** (page 38)
1. darkness
2. light
3. Jesus

**Fireflies** (page 41)
1. lightning bugs
2. invertebrate
3. grassy areas, marshes, places with water, or woods
4. smaller insects
5. carnivore
6. birds, other fireflies, and reptiles
7. bad-tasting substance in their blood
8. green, red, or yellow
9. a natural light coming from the bodies of some animals (fireflies and jellyfish)
10. when it is dark
11. males (female fireflies light the ground trying to attract male fireflies)

**Jellyfish** (page 43)
1. no (jellyfish do not have vertebrates)
2. invertebrates
3. carnivore
4. small fish and zooplankton
5. ocean currents
6. some fish, larger jellyfish, penguins, sea turtles, and sharks
7. tentacles containing venom
8. bioluminescence (natural light coming from the bodies of some animals- fireflies and jellyfish)
9. blue, lilac, orange, purple, red, and yellow (some are clear)
10. up to one year
11. 95% water

**Stalagtites and Stalagmites** (page 46)
1. cave
2. stalagmites
3. stalactites

**Brown Bears** (page 48)
1. cubs
2. grizzly bears
3. vertebrates
4. warm-blooded
5. omnivores
6. plants: berries, fruits, and nuts
animals: amphibians, fish, insects, mammals, and reptiles
7. forests and mountains
8. hibernate
9. birth cubs
10. cougars, humans, and wolves
11. claws, size, strength, and teeth

**Bat** (page 51)
1. pups
2. vertebrates
3. warm-blooded when awake; cold-blooded when asleep
4. caves, trees, bridges, mines, and other structures
5. carnivore or herbivore
6. some eat insects; others eat fruit; one species eats blood from other animals
7. echolocation: bouncing sounds off insects to locate them
8. hawks, owls, raccoons, and snakes
9. large numbers: colonies of 100 to 1,000 bats
10. bat

## Unit 2: Sky and Water

**Clouds** (page 57)
1. cirrus
2. nimbus
3. cumulus
4. cirrus
5. stratus
6. cumulus
7. stratus
8. nimbus

**Water Cycle** (page 63)
1. evaporation
2. condensation
3. precipitation

**The Invisible** (page 69)
1. invisible
2. atoms
3. microscope

**Basilisk Lizard** (page 72)
1. Jesus Christ lizard
2. hinges, hollow bones, and air-pockets on their feet
3. 15 mph; 33-66 feet before sinking
4. reptile
5. lays eggs in a trench
6. hatchlings
7. vertebrates
8. cold-blooded
9. tropical forests
10. omnivores
11. insects, smaller lizards, snails, spiders, and plants
12. birds, fish, nocturnal mammals, and snakes
13. quick speed and the ability to "walk" on water

**Float or Sink** (page 74)
1. floats
2. sinks

## Unit 3: Dry Land and Plants

**Plants** (page 82)
1. bud
2. flowers
3. leaves
4. seeds
5. stem
6. fruits and leaves
7. xylem
8. the process of pulling water from the soil into flowers and leaves

**The Sower and the Seed** (page 86)
1. plants
2. crops
3. soil
4. thorns

**Poisonous Plants : Matching** (page 91)
1. F
2. A
3. C
4. B
5. D
6. E

**Poisonous Plants** (page 92)
1. poison
2. poison ivy or poison oak
3. water hemlock
4. oleander
5. deadly nightshade

**Habitats: Matching** (page 97)
1. D
2. E
3. B

**Habitats** (page 99)
1. air, shelter, water, and food
2. three days
3. forest
4. ocean
5. poles
6. grasslands
7. fresh water
8. desert
9. answers will vary

## Unit 4: Sun, Moon, Stars, and Planets

**Sun and Other Stars** (page 108)
1. sun
2. sun
3. white (the atmosphere makes it appear orange)
4. dark images created when objects block light
5. astronomy: study of the universe; astrology: using constellations to predict events
6. moon and stars
7. to determine when to plant and harvest
8. eighty-eight
9. as a navigational tool (The two stars at the end of the Big Dipper point to the North Star.)

**Earth's Moon** (page 114)
1. twenty-eight days
2. moon, planet, or machine that orbits a planet or star
3. reflect
4. Neil Armstrong
5. first quarter
6. full moon
7. third quarter
8. new moon

**Planets** (page 118)
1. Mercury
2. Neptune
3. Saturn
4. Earth
5. Jupiter
6. Mars
7. eight (Pluto is now considered a dwarf planet.)
8. no
9. an instrument used to collect and magnify the light from distant objects

**Gravity and Magnetism** (page 122)
1. to increase rate or speed
2. Galileo Galilei
3. Isaac Newton
4. amount of matter in an object
5. force
6. gravity
7. land on the ground at exactly the same time
8. attract: pull towards; repel: push away

## Unit 5: Fish, Boats, Birds, and Airplanes

**Fish** (page 126)
1. over 3,000 species
2. vertebrates
3. cold-blooded
4. lay eggs
5. algae, amphibians, crustaceans, fish eggs, insects, mollusks, plankton, plants, and worms
6. no; they rest
7. no; both are mammals

**Sharks** (page 127)
1. medicines, vitamins, and shark meat
2. salt water habitat
3. spot pattern unique to each shark (similar to the idea that no human fingerprint is the same)
4. carnivore
5. other fish
6. can see in dark places; great night vision; can see colors
7. no natural predators except humans
8. great eyesight, fast swimming, and numerous sets of replaceable teeth

**Bluefin Tuna** (page 128)
1. highly-prized food
2. saltwater habitat
3. young bluefin and tiny zooplankton
4. billfishes, humans, sharks, and toothed whales,
5. thirty years or more
6. up to 10 million eggs

**Atlantic Salmon** (page 129)
1. born in freshwater, migrate to sea, and return to freshwater to reproduce
2. silver sheen with pockets of blue, crimson, purple, pink, or green most of the year; scales change to a rainbow of colors when reproducing
3. over 4,000 eggs
4. 3-126 pounds
5. crayfish, eels, grasshoppers, herring, krill, mayflies, shrimp, squid stoneflies, and zooplankton
6. bears, eagles, humans, otters, seals, and sharks

**Boats** (page 134)
1. stern; tail
2. keel; trunk
3. hull
4. scales
5. rudder; tail and fins
6. bow; head
7. sails
8. gills

**Birds** (page 137)
1. lay eggs
2. warm-blooded
3. vertebrates
4. beak
5. feathers
6. yes

**Eagle** (page 138)
1. eaglets
2. amphibians, crabs, fish, and reptiles
3. carnivores
4. no predators except humans
5. typically mate for life (If one dies, an eagle might find a new mate maintaining the same territory.)
6. Females take care of their young (males find food.)
7. high cliffs
8. An eagle's eyes are eight times stronger than a human's eyes.
9. cunning, bold, and intelligent

**Vulture** (page 139)
1. chicks
2. carnivores
3. dead things (rotten meat that is toxic to other animals)
4. very few predators (hawks or eagles may steal baby vultures from their nests)
5. Both the male and female take care of their young for about three months.
6. no (prevent disease by devouring dead animals)
7. can smell dead animals from a mile away
8. soaring
9. no proper nests (lay eggs in dense thickets, hollow logs, and caves)
10. yes

**Raven** (page 140)
1. chicks
2. omnivores
3. dead things, eggs, grains, insects, rodents, and worms
4. goshawks, hawks, and owls
5. Both parents take care of their young.
6. elaborate chases, dives, and rolls
7. yes

**Airplanes** (page 145)
1. lift, drag, gravity (weight), and thrust
2. someone who makes new things
3. Orville and Wilber Wright
4. lift
5. thrust
6. drag
7. gravity (weight)

## Unit 6: Animals and People

**Mammals** (page 150)
1. vertebrates
2. warm-blooded
3. most give birth to live young (exceptions: a spiny anteater and duck-billed platypus lay eggs)
4. carnivore
5. herbivore
6. omnivore
7. prey
8. predator
9. hair
10. nurse

**Wolf** (page 151)
1. mammal
2. pups
3. vertebrates
4. warm-blooded
5. deserts, forests, grasslands, mountains, and the tundra
6. carnivores
7. beaver, caribou, deer, elk, mice, moose, rabbit, and squirrel
8. top of the food chain with very few predators; main predator is humans.
9. sharp teeth and claws; hunt in packs
10. wild: around 5-12 years; zoos: up to 20 years.
11. tend to mate for life, unless one of them dies

**Sheep** (page 152)
1. mammal
2. lambs
3. vertebrates
4. warm-blooded
5. grasslands
6. herbivores
7. grass and vegetation
8. bears, bobcats, coyotes, dogs, eagles, lions, lynxes, wild pigs, and wolves
9. Llamas or mules are often mixed in with sheep herds to ward off predators; shepherds protect sheep.

**Arthropods** (page 154)
1. invertebrates
2. cold-blooded
3. everywhere
4. omnivores, herbivores, or carnivores, depending on the species
5. exoskeleton, segmented body parts, and jointed appendages; cold-blooded and most lay eggs
6. lay eggs
7. an arthropod's outer hard covering
8. shedding of the exoskeleton
9. wormlike arthropods with many legs (centipedes & millipedes)

10. sea arthropods with hard body coverings (lobsters and crabs)
11. arthropods with two body parts and eight legs
12. arthropods with six legs, exoskeleton, three body parts (head, thorax, and abdomen), two antennae, and wings

**Lobster** (page 155)
1. carnivores
2. fish, crabs, clams, mussels, sea urchins, and other lobsters
3. dog fish, skates, cod, and humans
4. hard outer shell and claws
5. grows back
6. most get eaten but can live to over 100 years old
7. by molting
8. crustaceans

**Butterflies and Moths** (page 157)
1. developmental changes (caterpillar to a butterfly; tadpole to a frog)
2. active in the day
3. moth
4. insects

**Centipedes and Millipedes** (page 158)
1. 15-30 legs
2. 47-197 legs
3. birds, toads, frogs, and small mammals
4. centipedes: small bugs, such as silverfish, spiders, and ants; millipedes: decaying matter, such as fruits, vegetables, and plants
5. centipedes (poisonous bite that feels as painful as a bee sting)
6. both grow new legs
7. myriapods

**Tarantula** (page 159)
1. eight
2. two body parts: head and abdomen
3. no
4. Pepsis wasp
5. grasshoppers, beetles, other small spiders, and sometimes small lizards
6. no
7. arachnids

**Reptiles** (page 162)
1. vertebrates
2. cold-blooded
3. most lay eggs
4. most are carnivores; some are herbivores
5. scales
6. Poison is delivered by swallowing, inhaling, or seeping through the skin; venom is injected through a bite or a sting.

**Sea Turtles and Tortoises** (page 166)
1. sea turtles: lakes, ocean, ponds, rivers, swamps, and wetlands; tortoises: deserts, grasslands, scrubs, and tropical forests
2. omnivores
3. sea turtles: algae, crabs, fish, jellyfish, seagrasses, seaweed, shrimp, and sponges;
tortoises: clover, fungi, grasses, millipedes, and mulberry leaves
4. sea turtles: humans and sharks (hatchlings die by birds, crabs, fish, and octopus);
tortoises: coyotes, foxes, ravens, and roadrunners
5. sea turtles: up to 100 years old; tortoises: up to 150 years and beyond
6. no (beaks)
7. sea turtles: sandy beaches; tortoises: dens, sandy areas, or under objects such as logs
8. hard shells

**Snakes** (page 167)
1. forked tongue
2. with its tongue
3. every continent except Antarctica
4. no (85% are not poisonous)
5. carnivores

6. amphibians, birds, fish, small mammals, and other snakes
7. birds, coyotes, foxes, mongooses, raccoons, and wild boars
8. camouflage, defensive coiling, and the ability to slither away
9. yes

### **Alligators and Crocodiles** (page 170)

1. strong jaws, swimming skills, and long powerful tails; excellent eyesight, hearing, and sense of smell
2. yes, but only for a short distance
3. carnivores
4. fish, deer, frogs, birds, and buffalo
5. adults: humans; hatchlings: birds, and crabs, and raccoons
6. nearly an hour
7. yes
8. alligators: freshwater; crocodiles: freshwater and saltwater
9. crocodiles are larger than alligators
10. alligator: snout is wide and in the shape of a U; crocodile: snout is slender and in the shape of a V

### **Frogs and Toads** (page 173)

1. tadpoles
2. carnivores
3. most eat insects, spiders, worms, and slugs; larger species eat mice, birds, other small reptiles and amphibians
4. long, sticky tongues
5. near or in the water
6. frogs: birds, foxes, and snakes; toads: birds, raccoons, and snakes
7. 7-14 years
8. frogs: long, powerful back legs – great for jumping; toads: short back legs
9. frogs: moist and smooth; toads: dry and bumpy
10. frogs: live mostly in water; toads: live mostly on land
11. no (beaks instead of teeth)

### **Salamanders** (page 175)

1. efts
2. carnivores
3. flies, insects, mosquito larvae, other salamanders, slugs, spiders, and worms
4. long sticky tongues
5. attached to stones or debris in underground freshwater crevices
6. raccoons, skunks, snakes, and turtles
7. eggs, larva with gill buds, larva with 2 legs, larva with 4 legs, adult

### **Blind** (page 181)

1. Braille alphabet
2. Louis Braille
3. unable to see certain colors

### **Taste and Smell** (page 192)

Note: other answers can be correct. Answers will vary.

1. unsweetened chocolate, coffee, and black tea
2. potato chips, crackers, and French fries
3. sauces, beef jerky, and meat
4. lemons, grapefruits, and limes
5. fruit, cookies, and honey
6. essential oils

### **Touch** (page 197)

1. loops
2. arches
3. whorls
4. no
5. no

## Unit 7: God Rested

### **Rest** (page 203)

1. answers vary
2. 12-15 hours
3. 8-10 hours
4. germs
5. immune system

# CREATION
## ANSWERS (10)

| | | | | | | | | | | | | | |
|---|---|---|---|---|---|---|---|---|---|---|---|---|---|
| N | A | J | W | A | T | S | H | L | A | G | R | O | T |
| O | P | B | U | K | Y | L | A | N | D | B | N | F | C |
| I | Z | L | V | U | O | T | S | L | M | S | L | A | F |
| T | I | C | A | R | M | O | O | N | K | K | E | D | P |
| A | F | D | E | N | D | E | C | H | S | Y | T | C | R |
| E | H | N | E | P | T | W | E | I | B | A | D | E | H |
| R | G | U | A | F | G | C | N | I | J | U | T | L | A |
| C | F | O | V | M | D | O | U | I | B | A | G | N | T |
| V | U | E | W | K | U | E | S | J | W | O | I | X | I |
| A | D | B | S | C | V | N | I | M | D | G | G | E | H |
| U | O | A | O | T | L | U | O | A | H | W | A | O | I |
| K | O | X | Y | N | A | M | P | T | S | I | M | E | O |
| Y | P | U | I | Y | I | R | A | T | X | A | O | U | E |
| A | N | I | M | A | L | A | Z | I | R | E | Q | H | I |

| | | |
|---|---|---|
| LAND | MOON | WATER |
| DAY | ANIMAL | CREATION |
| NIGHT | SKY | PLANT |
| SUN | STAR | MAN |

# PLANTS
## ANSWERS (81)

| | | | | | | | | | | | | | |
|---|---|---|---|---|---|---|---|---|---|---|---|---|---|
| D | H | T | D | S | K | V | G | R | Z | L | Y | F | R |
| A | U | V | E | E | L | W | H | S | A | M | Z | G | S |
| B | I | B | F | E | M | X | L | T | B | N | A | H | T |
| R | J | U | C | D | N | E | I | R | C | O | B | I | U |
| E | K | W | A | H | A | Y | J | O | D | P | X | J | V |
| T | I | U | R | F | O | Z | K | O | E | Q | Y | K | W |
| A | L | X | N | L | P | A | L | T | F | R | L | L | X |
| W | M | S | A | O | Q | B | M | U | G | S | E | M | Y |
| C | N | Y | T | W | R | C | N | V | T | T | M | N | N |
| D | O | Z | I | E | S | D | O | N | H | U | C | U | Z |
| E | P | A | O | R | M | E | A | W | I | V | S | O | A |
| F | Q | B | N | I | T | L | P | X | J | W | D | P | B |
| T | R | A | N | S | P | I | R | A | T | I | O | N | C |
| G | S | C | G | J | U | F | Q | Y | K | X | E | Q | D |

| | | |
|---|---|---|
| BUD | FRUIT | PLANT |
| SEED | STEM | CARNATION |
| LEAF | XYLEM | WATER |
| ROOT | TRANSPIRATION | SUN |

# SOLAR SYSTEM
## ANSWERS (117)

| | | | | | | | | | | | | | |
|---|---|---|---|---|---|---|---|---|---|---|---|---|---|
| A | P | B | O | N | M | J | C | A | G | D | I | E | S |
| B | K | C | S | B | I | A | S | Y | J | Z | G | H | O |
| A | J | R | L | K | H | W | X | U | H | G | R | F | L |
| Z | A | D | I | U | S | T | P | S | N | P | F | O | A |
| M | Y | V | J | T | R | I | Q | U | F | A | Q | N | R |
| X | M | W | O | N | T | A | E | V | A | O | R | L | S |
| U | I | E | U | E | M | U | J | P | C | N | K | U | Y |
| H | S | O | R | G | E | F | E | I | N | E | U | M | S |
| U | E | P | V | C | L | H | D | C | R | P | B | S | T |
| A | V | W | I | T | U | W | T | O | U | T | E | A | E |
| Y | X | E | S | Y | X | R | Z | R | T | U | L | K | M |
| T | R | Q | P | D | I | O | Y | N | A | N | M | O | A |
| S | U | N | E | V | N | M | L | G | S | E | O | I | J |
| R | Z | Q | U | H | S | T | A | R | B | N | K | A | C |

| | | |
|---|---|---|
| MARS | JUPITER | EARTH |
| VENUS | URANUS | NEPTUNE |
| MERCURY | SATURN | SUN |
| MOON | SOLAR SYSTEM | STAR |

# FISH
## ANSWERS (130)

| | | | | | | | | | | | | | |
|---|---|---|---|---|---|---|---|---|---|---|---|---|---|
| A | D | I | D | H | Q | A | E | F | T | B | E | H | S |
| T | E | H | J | I | R | B | I | M | A | C | G | I | C |
| B | U | T | K | E | S | S | F | N | Z | D | L | M | A |
| C | F | N | E | C | H | A | B | I | T | A | T | P | L |
| G | R | U | A | U | T | L | G | O | Y | F | Q | U | E |
| Q | O | V | F | D | U | T | H | P | T | S | T | Z | S |
| M | T | L | G | O | V | W | I | Q | U | Y | B | C | D |
| N | A | W | K | R | S | A | L | M | O | N | E | F | G |
| O | D | X | L | P | W | T | J | U | R | X | J | I | H |
| P | E | Y | M | E | X | E | K | V | T | W | K | L | M |
| S | R | Z | N | R | Y | R | L | S | X | V | N | O | P |
| G | P | A | O | J | A | C | M | I | G | R | A | T | E |
| G | R | B | P | H | Z | D | K | R | W | U | V | W | Y |
| E | S | C | S | R | E | T | A | W | H | S | E | R | F |

| | | |
|---|---|---|
| FISH | HABITAT | SALTWATER |
| SHARK | SALMON | TROUT |
| TUNA | EGGS | FRESHWATER |
| REPRODUCE | MIGRATE | PREDATOR |

# AMPHIBIANS
## ANSWERS (176)

| | | | | | | | | | | | | | |
|---|---|---|---|---|---|---|---|---|---|---|---|---|---|
| A | S | I | S | O | H | P | R | O | M | A | T | E | M |
| F | G | H | J | K | L | Z | X | G | C | V | B | N | P |
| Q | W | E | R | T | Y | U | I | I | O | P | C | R | V |
| B | N | M | A | S | D | L | F | H | G | J | E | K | L |
| W | E | F | R | O | G | R | T | I | E | Y | U | I | O |
| S | G | G | E | A | S | D | L | F | R | G | H | J | K |
| R | T | A | D | P | O | L | E | R | O | T | S | Y | U |
| O | H | G | N | F | S | D | S | A | V | P | G | O | I |
| T | J | K | A | M | P | H | I | B | I | A | N | L | P |
| A | S | A | M | W | E | R | T | U | N | Y | U | I | S |
| D | F | D | A | O | T | G | H | J | R | K | L | N | L |
| E | O | I | L | U | Y | T | R | E | A | W | X | C | L |
| R | P | M | A | N | B | V | C | X | C | A | S | D | I |
| P | R | T | S | Y | U | I | O | P | L | K | J | H | G |

| | | |
|---|---|---|
| SALAMANDER | FROG | TOAD |
| AMPHIBIAN | TADPOLE | EGGS |
| LUNGS | METAMORPHOSIS | PREDATOR |
| CARNIVORE | GILLS | PREY |

# Vocabulary

**Acceleration:** to increase rate or speed

**Amphibian:** an animal who lives on land and water (frogs, toads, and salamanders)

**Arachnid:** an animal with eight legs and two body parts (daddy longlegs, scorpions, and spiders)

**Arch:** fingerprints where lines begin on one side, go up, and exit on the opposite side

**Armstrong, Neil:** the first man to walk on the moon

**Arthropod:** a cold-blooded invertebrate with an exoskeleton, segmented body parts and jointed appendages; most lay eggs (arachnids, crustaceans, insects, and myriapods)

**Astrology:** using constellations to predict events

**Astronomy:** scientifically-based study of the universe

**Atom:** the smallest part of any element; invisible

**Attract:** to pull towards

**Beak:** the mouth of a bird

**Bioluminescence:** the light coming from the bodies of some plants and animals

**Bitter:** unsweet (coffee, unsweetened chocolate, and black tea)

**Blind:** unable to see

**Bow:** the front of a boat

**Braille:** an alphabet for the blind

**Braille, Louis:** invented an alphabet for the blind

**Bud:** the earliest stage of a flower or fruit

**Camouflage:** the ability to blend into backgrounds by colors, patterns, or textures

**Carnivore:** a meat-eating animal

**Cause:** when an object makes something happen

**Cave:** a natural, underground hole that is completely dark

**Chain reaction:** when one action causes another action

**Cirrus clouds:** look like horse tails; a sign of fair weather

**Clouds:** condensed water vapor floating in the sky

**Cold-blooded:** an animal that takes its body temperature from the environment

**Color blindness:** unable to see certain colors

**Compass:** an instrument with a magnetized pointer indicating north

**Condensation:** gas changing into liquid

**Constellation:** a group of stars identified according to pattern

**Cornerstone:** a large stone used in ancient times during the construction of a building

**Crops:** farm plants

**Crustacean:** a sea animal with hard coverings (lobsters and crabs)

**Cumulus clouds:** white, puffy clouds; a sign of good weather

**Darkness:** the lack of light

**Deaf:** unable to hear

**Density:** the measurement of how heavy something is compared to its size

**Desert:** land with very little rainfall

**Drag:** a force that opposes the motion of planes due to friction

**Echolocation:** bouncing sounds off insects to locate them

**Effect:** a consequence
**Essential Oils:** natural oils from plants (lavender, peppermint, and clove)
**Evaporation:** liquid turning into gas
**Ewe:** a female sheep
**Exoskeleton:** the hard covering that supports and protects arthropods
**Feathers:** the covering of a bird's body
**Fingerprints:** impressions of fingertips
**Fins:** the part of a fish used for movement and steering
**First Quarter:** a phase of the moon; looks like a D
**Fish:** cold-blooded water vertebrates who have gills, fins, and lay eggs
**Flippers:** the body parts on the sides of sea turtles that help with swimming
**Float:** when an object stays on top of the water and does not sink
**Flower:** the part of a plant with petals; produces fruits and seeds
**Force:** the push or pull of an object
**Forest:** land filled with trees
**Foundation:** the supporting materials underneath a building (stone or sand)
**Fresh Water:** water with little or no salt
**Fruit:** the edible part of a plant that contains seeds
**Full Moon:** a phase of the moon; looks like a white circle
**Galilei, Galileo:** a famous scientist who studied gravity
**Germs:** disease-causing microscopic organisms
**Gills:** the part fish and amphibians use to breathe oxygen from water
**Grassland:** land covered with different kinds of grasses
**Gravity:** an invisible force that pulls objects towards the center of the earth
**Habitat:** any place where plants and animals live and grow
**Head:** the front of a fish
**Herbivore:** a plant-eating animal
**Hibernate:** a deep sleep some animals take in order to get through the winter
**Horizontal:** east/west; left/right; where the earth appears to meet the sky (horizon)
**Hull:** the body of a boat
**Immune System:** a system of the body that fights disease
**Insect:** an animal with six legs, exoskeleton, three body parts, two antennae, and wings
**Inventor:** someone who makes new things
**Invisible:** something our eyes cannot see
**Invertebrate:** an animal without a backbone
**Keel:** lower structure of a boat from bow to stern
**Lamb:** a baby sheep
**Last Quarter:** a phase of the moon; looks like a C
**Leaf:** the flat, green part of a plant that grows from a stem
**Leprosy:** a disease resulting in the loss of the sense of touch
**Lift:** a force that opposes gravity and keeps planes in the air
**Light:** a type of energy allowing sight

**Loop:** fingerprints where some lines start on one side, curve upward, and go back around to the side where they began

**Lungs:** the organs used for breathing oxygen in the air

**Magnetism:** an invisible magnetic force observed when magnets attract or repel

**Mammal:** a warm-blooded vertebrate that has hair and nurses; most give birth to live young

**Mass:** the amount of matter in an object

**Mast:** a long pole that rises from the bottom of a ship and supports the sails and rigging

**Metamorphosis:** developmental changes (caterpillar to a butterfly; tadpole to a frog)

**Microscope:** an instrument used to make very small objects larger

**Molting:** shedding old feathers, hair, scales, or skin to make way for new growth

**Moon:** a planet's satellite

**Myriapod:** a wormlike animal with many legs (centipedes and millipedes)

**New Moon:** a phase of the moon when it cannot be seen

**Newton, Isaac:** the first person to define a theory of gravity

**Nimbus clouds:** dark clouds; rain or snow inside them

**Ocean:** salt water covering almost three-fourths of the earth's surface

**Omnivore:** a plant and meat-eating animal

**Orbit:** a repeating path one object takes around another object

**Planets:** large masses that orbit the sun

**Plants:** living things with seeds, roots, stems, leaves, and flowers

**Plumb:** vertical and straight

**Plumb line:** a line with a weight on one end used to determine if something is vertically straight

**Poison (a):** a substance found in plants that can cause illness or death

**Poison (b):** harmful chemicals delivered by swallowing, inhaling, or seeping through the skin

**Poles:** cold land covering the top and bottom of the earth (North and South Pole)

**Precipitation:** rain, snow, sleet, or hail falling to the ground

**Predator:** an animal hunting another animal

**Prey:** an animal hunted by another animal

**Primary colors:** red, yellow, and blue

**Pup:** a baby wolf

**Ram:** a male sheep

**Reflective light:** light bouncing off an object (glass, water, or polished metal)

**Refractive light:** light bending (magnifying glass, prism, or rainbow)

**Repel:** to push away

**Reptile:** a cold-blooded vertebrate that usually has scales and lays eggs

**Rings:** the circles around some planets made of rock, ice, and dust particles

**Roots:** the part of a plant that grows underground to absorb water and nutrients

**Rudder:** the part used by boats for steering

**Sabbath:** a day to worship and rest from work; kept by Jewish people from Friday evening to Saturday evening and by many Christians all day Sunday

**Sail:** the piece of fabric used to move a boat forward by wind

**Salty:** saline (potato chips, crackers, and French fries)

**Satellite:** a moon, planet, or machine orbiting a planet or a star

**Savory:** pleasant, full flavor; not sweet (sauces, beef jerky, and meat)

**Scales (a):** rigid plates that grow on the skin of a fish
**Scales (b):** plates made of keratin, like hair, and found in an overlapping arrangement on the skin of a snake
**Secondary colors:** orange, green, and purple
**Seeds:** the smallest part of a plant that grows into a new plant
**Shadows:** dark images created when objects block light
**Sign language:** communication with hands used by those who are deaf
**Sink:** when an object goes lower than the surface of the water
**Sleep:** the rest cycle
**Soil:** the top layer of the earth
**Solar System:** the eight planets and their moons in orbit around the sun
**Sour:** acidic, not sweet (lemons, grapefruits, and limes)
**Spider:** an arachnid with fangs that produce venom; often spins webs of silk to catch prey
**Stalactites:** dripping mineral water growing from cave ceilings into rock structures
**Stalagmites:** dripping mineral water growing from cave floors into rock structures
**Stars:** lights in the night sky
**Stem:** the part of a plant that supports branches and leaves
**Stern:** the back of a boat
**Stratus clouds:** flat sheets of clouds; fog; low to the ground
**Sun:** a star that gives the earth heat and light
**Sweet:** sugary (fruits, honey, and maple syrup)
**Tail:** the part at the end of a fish used for swimming and steering
**Telescope:** an instrument used to collect and magnify light from distant objects
**Thorns:** sharp points located on the stems of some plants
**Thrust:** a force that pushes planes through the air
**Transpiration:** the process of pulling water from the soil into flowers and leaves
**Trunk:** the middle part of a fish
**Venom:** harmful chemicals injected through a bite or a sting
**Vertebrate:** an animal with a backbone
**Vertical:** up/down; north/south
**Warm-blooded:** an animal that maintains a constant body temperature
**Water Cycle:** evaporation, condensation, and precipitation
**Whorl:** fingerprints with a circular pattern
**Wing:** the part of a bird used for flying
**Wright, Orville and Wilbur:** the brothers who invented the first airplane
**Xylem:** a tube in the stem that pulls water and nutrients upward from the root

# Exploring God's Creation
## Images and Sources

### Images
Earth: Adobe, 2

## Creation, Sin, and Rescue
### Images
Days of Creation: Jeanette Summerville, Guatemala Deaf Ministries, 5
Creation in God's Hand: Adobe, 6
The Fall: Jeanette Summerville, Guatemala Deaf Ministries, 7
Cain and Abel: Jeanette Summerville, Guatemala Deaf Ministries, 8
He is Risen: Jeanette Summerville, Guatemala Deaf Ministries, 9

### Sources
Genesis 2:16-17: New International Version, (NIV), 7
Genesis 4:6-7: New American Standard Bible (NASB), 8
I Thessalonians 4:16-18: New American Standard Bible (NASB), 9

## Chain Reactions
### Images
Chain Reaction: Adobe, 12
Dominoes: Shutterstock, 13

### Sources
Psalm 33:6, 9: New American Standard Bible (NASB), 12

## A Firm Foundation
### Images
Plumb-line Grandma: Jeanette Summerville, 14
House on the Rock: Adobe, 15
Wooden Blocks: Adobe, 16
Egyptian Plumb-line: Jeanette Summerville, 17
Plumb-line: Adobe, 18
Two Paths: Jeanette Summerville, Guatemala Deaf Ministries, 19
Cornerstone: Jeanette Summerville, 20
Three Little Pigs: Shutterstock, 21

### Sources
Luke 6:48: New American Standard Bible (NASB), 14
Matthew 7:24-27: New American Standard Bible (NASB), 15
Plumb Line History: Wonky Donkey Tools, 17-18
Isaiah 28:17: New International Version (NIV), 18
Matthew 17:13-14: New American Standard Bible (NASB), 19
Acts 4:11-12: New American Standard Bible (NASB), 20
Cornerstone History: Got Questions, 20

## Unit 1: Day and Night
### Images
Seeing Grandma: Jeanette Summerville, 22
Day and Night: Jeanette Summerville, Guatemala Deaf Ministries, 23

### Sources
Genesis 1:3-5: New American Standard Bible ((NASB), 23

## Reflective and Refractive Light
### Images
Reflective Light (swan): Adobe, 24
Refractive Light (gem): Adobe, 24
Reflective Light: Adobe, 25
Refractive Light: Jeanette Summerville, 26
Reflective and Refractive Light (circular rainbow): Adobe, 27

### Sources
I John 1:7: New International Version, (NIV), 24
Reflective and Refractive Light: Make Me Genius (YouTube), 24-27

## Colors
### Images
Building the Ark: Jeanette Summerville, Guatemala Deaf Ministries, 28
The Rainbow: Jeanette Summerville, Guatemala Deaf Ministries, 29
Rainbow: Jeanette Summerville, 30-31
Mixing Colors: Adobe, 33
Color Wheel: Jeanette Summerville, 34
Rainbow scene: Adobe, 35

### Sources
Genesis 9:13-16: New Internationl Version (NIV), 29
Rainbow Facts: How Stuff Works, 30-31
Psalm 52:8: New American Standard Bible (NASB), 32
Psalm 23:2: New American Standard Bible (NASB), 32
Matthew 28:3: New American Standard Bible (NASB), 32
Isaiah 1:18: New American Standard Bible (NASB), 32
Daniel 5:16: New American Standard Bible (NASB), 32
Matthew 2:11: New American Standard Bible (NASB), 32
Proverbs 25:11: New American Standard Bible (NASB), 32

## Light and Darkness
### Images
Sun and Moon: Adobe, 36
Light Bulb Worksheet: Jeanette Summerville, 37
Jesus is the Light: Adobe, 38
Empty Tomb: Adobe, 38
Firefly: Adobe, 39
Firefly: Shutterstock, 39
Firefly: Jeanette Summerville, 40
Fireflies: Shutterstock, 41
Jellyfish: Shutterstock, 42
Immortal Jellyfish: Adobe, 42
Jellyfish: Adobe, 43
God's Light: Adobe, 44
Man's Light: Adobe, 44

### Sources
Matthew 5:15-16: New American Standard Bible (NASB), 36
I John 1:9: New American Standard Bible (NASB), 37
John 8:12: New American Standard Bible (NASB), 38
Fun Facts About Fireflies: Bee Better Naturally, 39-41
Jellyfish: Sci Show Kids (YouTube), 42-43
Immortal Jellyfish: Natural History Museum, 42

## Caves, Brown Bears, and Bats

### Images
Stalactites and Stalagmites: Shutterstock, 45
Cave: Adobe, 46
Brown Bear Eating Fish: Shutterstock, 47
Brown Bear Eating Berries: Shutterstock, 47
Brown Bear Roaring: Shutterstock, 47
Brown Bear and Cubs: Adobe, 48
Brown Bear Art: Jeanette Summerville, 49
Bat Flying: Shutterstock, 50
Baby Bats: Shutterstock, 50
Bat: Jeanette Summerville, 51-52

### Sources
Genesis 1:3-5: New King James Version (NKJV), 45
Caves: National Geographic, 45-46
Brown Bear: Character Sketches Vol.1, 47-49
Bat: Character Sketches Vol.1; Biggest Book of Animals, 50-52

## Unit 2: Sky and Water

### Images
Scuba Diving Grandma: Jeanette Summerville, 53
Sky and Water: Jeanette Summerville, Guatemala Deaf Ministries, 54

### Sources
Genesis 1:20: New Internationl Version (NIV), 53
Genesis 1:6-8: New American Standard Bible (NASB), 54

## Clouds

### Images
Clouds: Shutterstock, 55
Cloud Worksheet: Jeanette Summerville, 56
Stratus, Cirrus, Cumulus and Nimbus Clouds: Guatemala, 57

### Sources
Job 35:5: New American Standard Bible (NASB), 55
Psalm 147:7-8: New King James Version (NKJV), 56
Clouds: Happy Learning (YouTube), 55-57

## The Water Cycle

### Images
Water Cycle: Jeanette Summerville, 58-59
Evaporation: Adobe, 60
Condensation: Pixabay, 61
Precipitation: Adobe, 62
Water Cycle: Shutterstock, 63

### Sources
Amos 9:6: New American Standard Bible (NASB), 58
Water Cycle: Hopscotch (YouTube), 58-63
Happy Learning (YouTube), 58-63

## The Invisible

### Images
Sound Waves: Pixabay, 64
Guitar: Pixabay, 65
Invisible Ink: Adobe, 66
Handwashing: Adobe, 67
Windmill: Adobe, 68
Boy and a Microscope: Adobe, 69

### Sources
Timothy 1:17: King James Version (KJV), 64

## Float or Sink

### Images
Jesus Walking on Water: Shutterstock, 70
Basilisk Lizard: Adobe, 71
Basilisk Lizard: Shutterstock, 72
Water Tension: McAuliffe-Shepard Discovery Center, 73
Floating Egg: Adobe, 74
Floating Plants: Pixabay, 75

### Sources
Matthew 14:22-33: New International Version (NIV), 70
Basilisk Lizard: National Geographic, AZ Animals, 71-72
Psalm 133:1: King James Version (KJV), 73

## Unit 3: Dry Land and Plants

### Images
Garden Grandma: Jeanette Summerville, 76
Land and Plants: Jeanette Summerville, Guatemala Deaf Ministries, 77

### Sources
Matthew 13:23: New American Standard Bible (NASB, 76
Genesis 1:11-13: New Internationl Version (NIV), 77

## Plants

### Images
Parts of Plants: Jeanette Summerville, 78-79
Transpiration: Adobe, 80
Flowers: Adobe, 82

### Sources
Ecclesiastes 3:1-2: New American Standard Bible, (NASB), 78
John 12:24: New American Standard Bible (NASB), 78

## The Sower and the Seed

### Images
The Sower and the Seed: Adobe, 83
Good and bad soil compared: Adobe, 84
Grass: Adobe, 85
Farm family: Adobe, 86

### Sources
Matthew 13:3-9: New American Standard Bible (NASB), 83
Matthew 13:19-23: New American Standard Bible (NASB), 83

## Poisonous Plants

### Images
Deadly Nightshade: Shutterstock, 87
Water Hemlock: Pixabay, 88; 91
Oleander: Pixabay, 88; 91
Daffodil: Pixabay, 89; 91
Deadly Nightshade: Shutterstock, 89; 91
Poison Oak: Shutterstock, 90; 91
Poison Ivy: Pixabay, 90; 91
Mushroom: Pixabay, 92

### Sources
Deuteronomy 29:18b: New American Standard Bible (NASB), 87
Poisonous Plants: World Atlas; Britannica, 87-92

## Habitats

### Images
Survival Rules: Shutterstock, 93
Grasslands: Adobe, 94; 97
Forest: Adobe, 94; 97
Poles: Adobe, 95; 97
Desert: Adobe, 95; 97
Fresh Water: Adobe, 96; 97
Salt Water: Adobe, 96; 97
Farm Habitat: Adobe, 98
Habitats: Shutterstock, 99

### Sources
Habitat Song: Hopscotch (YouTube), 93-99
John 1:3: New International Version (NIV), 93
USA Grassland Animals: National Park Service, 94; 97
USA Grassland Plants: Sciencing, 94; 97
USA Forest Animals: Tree farm system, 94; 97
USA Forest Plants: EEK Wisconsin, 94; 97
Pole Animals (Alaska): Alaska Department of Fish and Game, 95; 97
USA Poles Plants: Go Weather Forecast, 95; 97
USA Desert Animals: Desert USA, 95; 97
USA Desert Plants: PBS Green Planet, 95; 97
USA Fresh Water Animals: Wildlife Informer, 96; 97
USA Fresh Water Plants: National Geographic, 96; 97
Salt Water Animals: Lonely Planet, 96; 97
Salt Water Plants: Scuba, 96; 97

## Unit 4: Sun, Moon, Stars, and Planets

### Images
Star Gazing Grandma: Jeanette Summerville, 100
Sun, Moon, and Stars: Jeanette Summerville, Guatemala Deaf Ministries, 101

### Sources
Genesis 1:14-19: New Internationl Version (NIV), 101

## Sun and Other Stars

### Images
Sun: Adobe, 102
Ursa Minor and Ursa Major: Adobe, 103
Orion: Shutterstock, 104
Pleiades: Shutterstock, 104
Boy Shadow: Shutterstock, 105
Hand Shadows: Shutterstock, 106
Shadows Dancing: Adobe, 108

### Sources
Jeremiah 31:35: New King James Version (NKJV), 102
Revelation 22:5: New International Version (NIV), 102
Psalm 147:4: New American Standard Bible (NASB), 103
Job 9:9: New American Standard Bible (NASB), 103
Job 38:32: New American Standard Bible (NASB), 103
Difference Between Astronomy and Astrology: Scientific America 2020, 103

## Earth's Moon

### Images
Moon stamp (1969): Adobe, 109
Neil Armstrong: Shutterstock, 110
Moon Phases: Jeanette Summerville, 111; 113
Full Moon: Pixabay, 114

### Sources
Genesis 1:16-18: New American Standard Bible (NASB), 109
Neil Armstrong: Famous Scientists/Astronauts, 112
Lunar Phases: Sciencing, 109-114
Ephesians 5:1, English Standard Version (ESV), 110; 114

## Planets

### Images
Telescope: Shutterstock, 115
Solar System: Shutterstock, 118

### Sources
Hebrews 11:3: New International Version (NIV), 115
Planets: Hopscotch (YouTube), NASA, 115-118

### Sources
Habitat Song: Hopscotch (YouTube), 93-99
John 1:3: New International Version (NIV), 93
USA Grassland Animals: National Park Service, 94; 97
USA Grassland Plants: Sciencing, 94; 97
USA Forest Animals: Tree farm system, 94; 97
USA Forest Plants: EEK Wisconsin, 94; 97
Pole Animals (Alaska): Alaska Department of Fish and Game, 95; 97
USA Poles Plants: Go Weather Forecast, 95; 97
USA Desert Animals: Desert USA, 95; 97
USA Desert Plants: PBS Green Planet, 95; 97
USA Fresh Water Animals: Wildlife Informer, 96; 97
USA Fresh Water Plants: National Geographic, 96; 97
Salt Water Animals: Lonely Planet, 96; 97
Salt Water Plants: Scuba, 96; 97

## Unit 4: Sun, Moon, Stars, and Planets

### Images
Star Gazing Grandma: Jeanette Summerville, 100
Sun, Moon, and Stars: Jeanette Summerville, Guatemala Deaf Ministries, 101

### Sources
Genesis 1:14-19: New Internationl Version (NIV), 101

## Sun and Other Stars

### Images
Sun: Adobe, 102
Ursa Minor and Ursa Major: Adobe, 103
Orion: Shutterstock, 104
Pleiades: Shutterstock, 104
Boy Shadow: Shutterstock, 105
Hand Shadows: Shutterstock, 106
Shadows Dancing: Adobe, 108

### Sources
Jeremiah 31:35: New King James Version (NKJV), 102
Revelation 22:5: New International Version (NIV), 102
Psalm 147:4: New American Standard Bible (NASB), 103
Job 9:9: New American Standard Bible (NASB), 103
Job 38:32: New American Standard Bible (NASB), 103
Difference Between Astronomy and Astrology: Scientific America 2020, 103

## Earth's Moon

### Images
Moon stamp (1969): Adobe, 109
Neil Armstrong: Shutterstock, 110
Moon Phases: Jeanette Summerville, 111; 113
Full Moon: Pixabay, 114

### Sources
Genesis 1:16-18: New American Standard Bible (NASB), 109
Neil Armstrong: Famous Scientists/Astronauts, 112
Lunar Phases: Sciencing, 109-114
Ephesians 5:1, English Standard Version (ESV), 110; 114

## Planets

### Images
Telescope: Shutterstock, 115
Solar System: Shutterstock, 118

### Sources
Hebrews 11:3: New International Version (NIV), 115
Planets: Hopscotch (YouTube), NASA, 115-118

### Sources
Matthew 6:26: New American Standard Bible (NASB), 135
Isaiah 40:31: New American Standard Bible (NASB), 138
Eagles: Fun Facts. 138
Leviticus 11:13: New International Version (NIV), 139
Vulture: Spruce, 139
Genesis 8:6-7: New American Standard Bible (NASB), 140
Raven: National Geographic, 140

## Airplanes
### Images
Aircraft Motion: Shutterstock, 141
Wright Brothers' Plane: Adobe, 142
Wright Brothers' Stamp: Adobe, 142
787 Airplane: Shutterstock, 142
Floating Ping Pong Ball: Adobe, 143
Old Airplane: Adobe, 144
Airport Runway: Pixabay, 145
Paper Airplane: Pixabay, 146

### Sources
Wright and Flight: Bernoulli's Principle, EAA The Spirit of Aviation, 141-146

## Unit 6: Animals and People
### Images
Horse Riding Grandma: Jeanette Summerville, 147
Man and Animals: Jeanette Summerville, Guatemala Deaf Ministries, 148

### Sources
Genesis 1:24-25, 27, 31: New Internationl Version (NIV), 148

## Mammals
### Images
Mammals: Adobe, 149
Duck-billed Platypus: Shutterstock, 150
Spiny Anteater: Shutterstock, 150
Wolves: Shutterstock, 151
Sheep: Adobe, 152

### Sources
Genesis 1:24: New American Standard Bible, (NASB), 149
Isaiah 11:6: New American Standard Bible, (NASB), 151
Wolves: National Geographic; Wolf Haven International, 151
Genesis 29:3: New American Standard Bible, (NASB), 152
Sheep: Good Heart Animal Sanctuaries; Homestead Geek; Ontario Sheep, 152

## Arthropod
### Images
Spiderweb: Adobe, 153
Arthropods: Adobe, 154
Lobsters: Pixabay, 155
Moths and Butterflies Compared: Adobe, 156
Life Cycle: Jeanette Summerville, 157
Monarch Butterfly: Adobe, 157
Lepidoptera Moth: Adobe, 157
Centipede: Pixabay, 158
Millipede: Pixabay, 158
Tarantula: Pixabay, 159
Tarantula art: Jeanette Summerville, 160

### Sources
Leviticus 11:22: New American Standard Bible, (NASB), 153
Arthropods: Britannica; Merriam Webster, 153-160
Lobster: National Oceanic and Atmospheric Administration, 155
Matthew 6:20: New American Standard Bible, (NASB), 156
Moths and Butterflies: Live Science, 156-157
Centipedes and Millipedes: Cool Kid Facts, 158
Tarantula: National Geographic, 159-160

## Reptiles

### Images

Reptiles: Adobe, 161
Boa: Pixabay, 162
Anaconda: Adobe, 162
Rattlesnake: Pixabay, 162
Sea Turtle: Unsplash, 163
Tortoise: Adobe, 163
Sea Turtle: Jeanette Summerville, 164
Tortoise: Jeanette Summerville, 165
Baby Turtles Hatching: Shutterstock, 166
Corn Scaleless Snake: Shutterstock, 167
Crocodile art: Jeanette Summerville, 169
Baby Crocodile: Shutterstock, 170

### Sources

James 3:7: New American Standard Bible, (NASB), 161
Reptiles: Australian Academy of Science, One Kind Planet, 161-170
Reptiles: Everything Reptiles; Live Science; Ducksters; Everglades; National Park Service, 161-170
Sea Turtles: Sea Turtle Preservation Society, Turtle Foundation, A-Z Animals, 163-166
Snakes: Homeschool Pop; A-Z Animals; Wild Life Removal; California Herpes, National Geographic, 167

## Amphibians

### Images

Life Cycle of a Frog: Adobe, 171
Frog: Shutterstock, 172
Toad: Shutterstock, 172
Tadpoles: Shutterstock, 173
Salamander Life Cycle: Shutterstock, 174
Salamander: Shutterstock, 175

### Sources

Exodus 8:1-2: New American Standard Bible, (NASB), 171
Frogs and Toads: San Diego Zoo Wild Life Alliance; A-Z Animals, 171-173
Salamander: San Diego Zoo; Mass Audubon; Live Science, 174-175

## See

### Images

Guide Dog: Shutterstock, 177
Louis Braille: Shutterstock, 178
Braille Alphabet: Jeanette Summerville, 179
Reading Braille: Shutterstock, 179
Color Blind Test: Adobe, 180
Seeing Grandma: Jeanette Summerville, 181

### Sources

Matthew 9:27-29: New American Standard Bible, (NASB), 177
Matthew 6:23-24: New American Standard Bible, (NASB), 177
Louis Braille: Biography, 178-179
Colorblindness: National Eye Institute, 180

## Hear

### Images

Sign Language Alphabet: Shutterstock and Jeanette Summerville, 182
Listening: Pixabay, 183
Sign Language: Adobe, 184
Ten Commandments: Shutterstock, 185
Ten Commandments: Jeanette Summerville, Guatemala Deaf Ministries, 186

### Source

Mark 7:32-35: New American Standard Bible, (NASB), 182

## Taste and Smell

### Images

Taste and Smell: Adobe, 187
Essential Oils: Adobe, 189
Peppermint Oil: Adobe, 191
Lavender Field: Pixabay, 192

### Sources
Psalm 34:8: New American Standard Bible, (NASB), 187
Exodus 30:25: New American Standard Bible, (NASB), 187

## Touch
### Images
Fingerprint Detective: Pixabay, 194
Fingerprints: Shutterstock, 195
Fingerprint Art: Shutterstock, 196
Fingerprints: Shutterstock, 197

### Sources
Psalm 139:14: New American Standard Bible, (NASB), 194
Mark 1:41-42: New American Standard Bible, (NASB), 194
Fingerprints: Forensic Science Simplified, 194-197

## Unit 7: God Rested
### Images
Grandma Resting: Jeanette Summerville, 198
God Rested: Jeanette Summerville, Guatemala Deaf Ministries, 199

### Sources
Psalm 4:8: New American Standard Bible, (NASB), 198
Genesis 2:1-3: New American Standard Bible, (NASB), 199

## Sleep
### Images
Happy Face ZZZ, Adobe, 200
Healthy Sleep, Adobe, 201
Cat and Dog Sleeping, Shutterstock, 202
Baby Sleeping: Pixabay, 203
Dog Sleeping: Shutterstock, 204

### Sources
Mark 2:27: New American Standard Bible, (NASB), 200
Sleep Benefits: Office of Disease Prevention and Health Promotion, 200
Sleep Noise Colors: Amerisleep, 202
Sleep Questions: Better Kids, 202
Recommended Sleep: National Sleep Foundation, 203-204

## Finale Review

### Images
Chef Grandma: Jeanette Summerville, 205
Party Balloons: Adobe, 206

### Sources
Genesis 1:1: New Internationl Version (NIV), 205

www.ingramcontent.com/pod-product-compliance
Lightning Source LLC
LaVergne TN
LVHW060033210926
848481LV00013B/3854
*9798994893173*